ANODIZED!

ANODIZED!

Brilliant Colors & Bold Designs
FOR ALUMINUM
JEWELRY

Clare Stiles

LARK
CRAFTS
A Division of Sterling Publishing Co., Inc.
New York / London

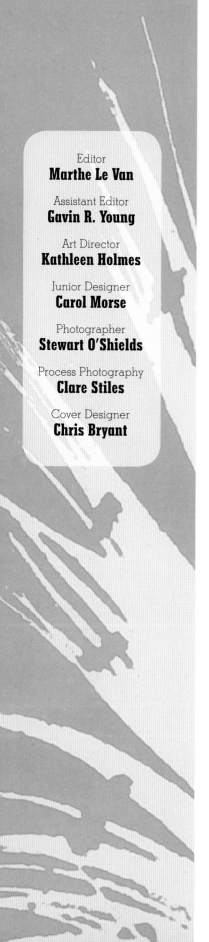

Editor
Marthe Le Van

Assistant Editor
Gavin R. Young

Art Director
Kathleen Holmes

Junior Designer
Carol Morse

Photographer
Stewart O'Shields

Process Photography
Clare Stiles

Cover Designer
Chris Bryant

Library of Congress Cataloging-in-Publication Data

Stiles, Clare.
  Anodized! : brilliant colors & bold designs for aluminum jewelry / Clare Stiles. -- 1st ed
    p. cm.
  Includes bibliographical references and index.
  ISBN 978-1-60059-520-2 (pbk. : alk. paper)
  1. Jewelry making. 2. Art metal-work. 3. Aluminum--Anodic oxidation. 4. Aluminum--Coloring. I. Title.
  TT212.S75 2010
  739.27--dc22

                              2010007073

10 9 8 7 6 5 4 3 2 1

First Edition

Published by Lark Books, A Division of
Sterling Publishing Co., Inc.
387 Park Avenue South, New York, NY 10016

Text © 2010, Clare Stiles
Photography © 2010, Lark Books unless otherwise specified
Illustrations © 2010, Lark Books

Distributed in Canada by Sterling Publishing,
c/o Canadian Manda Group, 165 Dufferin Street
Toronto, Ontario, Canada M6K 3H6

Distributed in the United Kingdom by GMC Distribution Services, Castle Place, 166 High
Street, Lewes, East Sussex, England BN7 1XU

Distributed in Australia by Capricorn Link (Australia) Pty Ltd., P.O. Box 704, Windsor, NSW
2756 Australia

If you have questions or comments about this book, please contact:
Lark Books
67 Broadway
Asheville, NC 28801
828-253-0467

Manufactured in China

ISBN 13: 978-1-60059-520-2

For information about custom editions, special sales, and premium and corporate
purchases, please contact the Sterling Special Sales Department at 800-805-5489 or
specialsales@sterlingpub.com.

For information about desk and examination copies available to college and university
professors, requests must be submitted to academic@larkbooks.com. Our complete policy
can be found at www.larkbooks.com.

Contents

INTRODUCTION

As THE DAUGHTER of a chartered mechanical engineer and theatrical costumier who made all my clothes from curtain fabric, maybe the path I have taken so far is no surprise.

Childhood memories of holidays at my grandparents' caravan at the English seaside—furnished with 1950s fabrics and candy-colored, polka-dot dishware—and visits to butterfly farms, aquariums, botanical gardens, and bird sanctuaries all form that huge part of me that is passionate about color.

I was introduced to anodizing during my first year of college when I was drawn like a magpie to a library book with a shiny red metallic spine. *Artists Anodizing Aluminum: The Sulfuric Acid Process*

Clare Stiles
Forged Rings and Bangles, 1998
From 2 to 7 cm
Aluminum; forged, anodized,
hand brush-blended
Photo by artist

Clare Stiles
Butterfly and Butterfly Mini Cuffs, 2004
4 x 6 cm and 5 x 6 cm
Aluminum; anodized, silkscreened,
submersion dyed
Photo by artist

by David LaPlantz, became a staple read for the next few years and the basis for all of my work as a professional maker. During my degree in three-dimensional design in metal, acclaimed jeweler Jane Adam visited, teaching us dyeing techniques and how to work with this incredible material. I was immediately fascinated by combining color with metal, which lead me away from my original intention to train as a blacksmith. I decided to combine my love of forging and color by cold-forging aluminum and anodizing it, still a unique process combination today. My final projects included extravagant costume headwear and jewelry.

After finishing college, I moved into a studio and started to supply shops and galleries with a wearable range inspired by the extraordinary colors in nature. After an accident in 2002, two broken hands lead to a change in my work. I reduced the amount of hammering necessary to complete a piece. Respected printmaker and good friend Bex Burchell re-introduced me to silkscreening and other printmaking techniques, which I adapted accordingly to use with anodized aluminum. My love of pattern and image was revisited, and it still forms the basis of my work. Making and teaching professionally since this time, I still haven't even scratched the surface of my ideas using this incredible material.

Jane Adam
Flower Pendants, 2006
Right pendant, 12 cm
Aluminum, silver, 9-karat gold;
anodized, dyed, crazed
Photo by Joël Degen

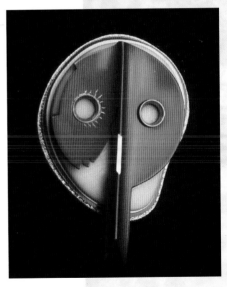

David LaPlantz
My Brave Face Brooch, 2000
6.6 x 5.5 x 6.6 cm
Aluminum, vegetable ivory, brass,
industrially painted aluminum;
anodized, dyed, riveted
Photo by artist

THE ART (& SIMPLE SCIENCE!) OF ANODIZING

This how-to book shows the creative use of anodized aluminum in jewelry. It looks at the anodizing, dyeing, printing, and coloring of the material, in addition to exploring skills including texturing, forming, and cold connection techniques. Presented alongside the instructional information are the very best examples of contemporary jewelers and metalsmiths using anodized aluminum. The skills and techniques presented in this book are drawn from textile design, watercolor painting, graphic design, and traditional printmaking. I carefully designed step-by-step projects so makers with little or no prior knowledge or experience can successfully complete pieces and so experienced makers can learn and expand their ideas to suit their individual needs. Whoever you are, whatever your experience level, I really hope you like it.

Britta L. Tobias
Growth, 2008
3.2 x 3.8 x 3.8 cm
Sterling silver, aluminum, fine silver, peridot, yellow cubic zirconia; anodized
Photo by artist

My intention is to give you an overview of anodized aluminum and provide the necessary information in a clear and concise manner without too much baffling scientific and chemical babble. My goal is for you to have a good understanding of the process for artistic purposes. Understandably, a large percentage of designers and makers use pre-anodized and ready-to-dye aluminum sheet bought from suppliers, mainly because of the chemicals and expensive equipment needed for anodizing. Some jewelry departments in colleges have anodizing baths and teach the process, but only a small percentage of students go on to employ the entire method. The majority use pre-anodized sheet. From my experience as a teacher, I feel that learning the actual anodizing process is an important part of understanding the material. Although not necessary, it is helpful, so the following text will provide a general overview. The techniques and projects in the remainder of the book will mainly focus on the use of pre-anodized sheet.

What Is Anodizing?

Aluminum is a very light and very soft white metal, making it perfect for industrial extrusion and die-casting. In simple terms, anodizing is an electrochemical process that creates a microscopic honeycomb-like pore on the surface of aluminum that absorbs dyes, creating a colored surface (figure 1). This anodic film hardens the surface of the metal, allowing components to be lightweight and hard wearing.

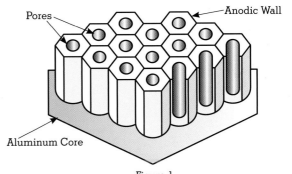

Figure 1
Microscopic view of the formation of the anodic film

Most industrial applications for anodizing aluminum are for corrosion resistance, hardening, and of course, coloring. Unlike other metal surface preparations, such as painting, powder coating, patination, and plating, an anodic film will not peel, crack, or corrode. This led NASA to use anodized aluminum for many components on satellites, shuttles, and space stations. There are numerous practical uses for this material in engineering, architecture, cookware, furniture, motor vehicles, aircraft, and boats to name but a few, and examples surround us everyday. Creative uses for anodized aluminum include sculpture, two- and three-dimensional design, product design, and of course jewelry. (One anodized and dyed product we can all recognize is the iPod with its multi-colored covers.)

A Bit of History

Chromic acid anodizing was invented in 1923 by Bengough-Stuart, an English company, for use by the British defense industry to protect seaplanes from corrosion. A few years later, in 1927, Gower and O'Brien patented the first sulfuric acid anodizing process. This is still the most commonly used method and the method that will be focused on throughout this book. Over the years many variations of anodizing have been mastered for different uses. Alternatives include organic, phosphoric, borate and tartrate baths, and plasma electrolytic oxidation. Other metals anodized include zinc, tantalum, niobium, and titanium.

A Happy Accident

Legend has it that years ago, when anodizing was used purely for industrial applications, no one knew that an anodized aluminum surface could be dyed. That was until spilled tea was found to permanently stain manufactured tea trays. It was discovered that these stains were actually dyed into the surface—and voila—coloring was introduced.

"No, It's Not Titanium..."

Because of my use of color, customers viewing my artwork often assume it's made from titanium—an easy mistake for the untrained eye. In fact, the difference between titanium and aluminum is substantial. Titanium, like aluminum, is a very light metal that can be anodized. Anodizing titanium, however, requires different processes and results in very different effects. Different, too, are the qualities and uses of anodized titanium. Titanium is a very hard material to begin with. The actual anodizing process *is* its coloring method. Unlike aluminum, oxidization on a titanium surface causes colors to reflect from the surface without the use of dyes. Anodized titanium's color range and type is very different from aluminum and less adaptable.

Poppy Porter
Racing Lace Cog Amulet, 2009
Necklace, 40 cm; pendant, 6 cm
Titanium, silver leaf, sterling silver, photograph, resin, silk cord; anodized; photo inlay, laser cut, forged
Photo by Angela Chan

Melody Armstrong
Ficus - Pendant, 2008
5.2 x 4.2 x 1.1 cm
Sterling silver, patina, titanium, citrine; anodized, riveted, hammered, pierced
Photo by artist

The Anodizing Process

The aluminum to be anodized is hung in a bath of a sulfuric acid/water electrolyte (15 to 18% sulfuric acid) forming the anode (the positive electrode).

The anodized film is "grown" by passing a direct current (DC) through the electrolyte releasing hydrogen at the cathode (the negative electrode).

Oxygen is produced at the aluminum anode causing a build up of aluminum oxide (figure 2).

The acid solution causes the aluminum to slowly dissolve the aluminum oxide. This, balanced with the oxidization from the pore-like surface, allows the electrolyte and current to reach the core of the aluminum and continue growing to a greater thickness (figure 3).

Process eats inward then grows outward to one third the depth.

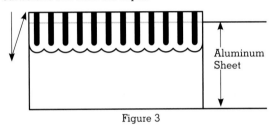

Figure 3

Anodizing begins by eating into the core of the metal then growing a porous film to a greater depth.

The resulting anodic film is measured in microns. The thickness of the film can be built up, and various thicknesses lend themselves to different particular uses.

Confused? If, like me, you're not very scientifically minded, don't panic. Really, unless you decide you would like to have a go at anodizing or set up your own anodizing unit, all you need to know is in figure 1 on page 9!

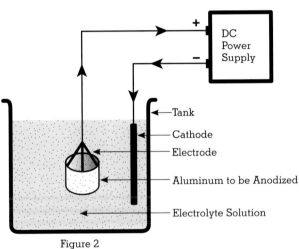

Figure 2
A typical anodizing setup

I-Ting Wang
Blooming III, 2004
11 x 4.5 x 2 cm
Aluminum, sterling silver,
cubic zirconia; dip dyed, photo
etched, hand painted, riveted
Photo by artist

How To Anodize

Anodizing aluminum can be carried out on many scales, from a garage workshop to an enormous factory. The equipment needed can range from a simple can of cola, a battery, and small DIY tanks with a car battery, to huge industrial facilities. To set up your own anodizing unit, costs can range from just a few dollars to many thousands. A typical small-scale setup that produces good results consists of a 20-liter bath connected to a Direct Current (DC) power supply (see photo, below). This is built in a fume cabinet with space for mixing the chemicals. Also, you'll want convenient access to a water supply for the rinsing stages.

Small-scale anodizing setup with a 20-liter bath

Health & Safety

If you are planning to try your hand at anodizing, I can't stress enough the importance of the following safety measures, and please research the process further. As I have said before, the information I am giving you is only an overview of the process.

Working with Chemicals

Sulfuric acid, caustic soda, and nitric acid are required to anodize your own aluminum. All of these acids give off hydrogen gases that can be combustible without proper ventilation. A fume cabinet or strong extraction device and fume masks and respirators are essential for the sake of your health and the health of others. Make sure your respirator contains a filter suitable for the chemicals you are working with and that it's not out-of-date. A respirator will not work if these requirements are not met. Equally important is the use of a chemical visor or chemical splash goggles (no vent holes!), suitable chemical resistant gloves, and protective or rubber boots. Baking soda (sodium bicarbonate) is an important safety material when working with chemicals, as it will neutralize acid in the event of an accident or spillage, so keep plenty close to hand.

When working with any chemicals, carefully read all supporting safety data well in advance of starting, and then read it again. It is also preferable to have had professional instruction or to work with a partner. Each of these situations will make for safer working conditions.

Right: **Jon M. Ryan**
Cylinder Ring, 2008
3.9 x 3 x 1.2 cm
Aluminum, sterling silver; carved,
fabricated, anodized
Photo by artist

Far right: **Ingeborg Vandamme**
Lovers Eye Necklace, 2008
11 x 11 x 1 cm
Aluminum, silver, glass, rubber, paper,
thread, ribbon; anodized
Photo by Francis Willemstijn

The Acid Rule

When mixing chemicals, you must remember the golden rules of the **three As**

ALWAYS ADD ACID to water—
never water to acid.

Forgetting this rule could result in you receiving very serious injury through chemical reaction or explosion.

Anodizing & the Environment

With the increase of environmental awareness, you may be questioning the friendliness of anodizing with chemicals. Anodizing uses water-based acids that are easily treated and safely disposable. The by-product (made up of aluminum hydroxide, aluminum sulfate, and water) actually can be of benefit at sewerage treatment plants, being used as filters at the secondary stage of treatment. The U.S. Environmental Protection Agency, which heavily regulates the use of chemicals, considers anodizing an environmentally friendly process. Please always check the safe disposal of chemicals by contacting your local authorities and following their advice.

Preparing the Metal

Preparation of the metal is key to successful anodizing. A chemically clean surface is essential for the anodizing process to work. To prevent the passing of any grease from your hands to the metal, you will not be handling the piece directly. Instead, hold the edges of the metal with clean gloves or brass or plastic tongs. Your workspace and tongs will need to be very clean and free from dust. Take care not to scratch or scuff the metal surface when placing it in the jig.

Preparing the Jig

A jig is a frame or clamp used to hang the metal in the bath. As shown in figure 4, simple jigs can be made from wire or old aluminum knitting needles. Large industrial jigs are specially designed so multiple pieces can be hung (see figure 5, page 14). The connection between the jig and the metal needs to be tight to ensure a current flow, so make sure the jig holds securely.

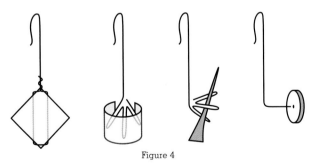

Figure 4

Examples of jigs and jigging differently shaped objects

Li-Chuan Lin
Internal Organs, 2006
Stomach: 16 x 15 x 3 cm
Liver: 9 x 17 x 3 cm
Heart: 8 x 14 x 8 cm
Aluminum, copper, ink; dip dyed, chasing & repoussé, riveted
Photo by artist

Another point to take into consideration is jig marks. Sometimes the connection point between the jig and the metal will not anodize. When jigging metal, bear this in mind, and keep the connection points small and in places that will not be immediately obvious. I tend to use 2- to 4-mm round wire to make a jig so the points of connection are small.

When making a cuff, I make my jig from wire with a hook for hanging and prongs spread out in three directions (see photo 1). I then squeeze the metal over the spread prongs, using the natural spring of the metal so the jig pushes outwards and holds the cuff securely.

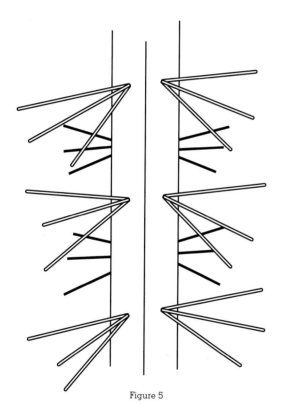

Figure 5

A section of an industrial jig for multiple pieces

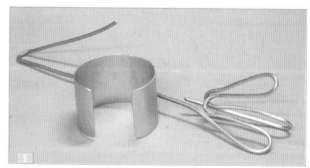

Aluminum cuff with wire jig

Niamh Mulligan
The King, 2008
12 x 12 x 1.5 cm
Sterling silver, anodized
aluminum; hand fabricated
Photo by Norton Associates

Cleaning the Aluminum

1. Select two chemical trays or old ovenproof dishes suitable in size for the aluminum you need to clean.

2. Determine how much fluid you will need to immerse the aluminum.

3. Mix the following two solutions at room temperature. Note: When mixing chemicals, distilled water is preferable, but I have never come across problems when using ordinary water.

   - Solution 1 - caustic soda/sodium hydroxide and water (follow amounts on bottle)

   - Solution 2 - between 10% and 30% nitric acid to between 70% and 90% water

4. Using detergent, thoroughly wash the aluminum.

5. Rinse the aluminum under running tap water for 1 minute.

6. Immerse the aluminum in the caustic etch (Solution 1) for 1 minute. You will see the solution working as a dark grey coating/smut starts to appear.

7. Rinse the aluminum under running tap water for 1 minute.

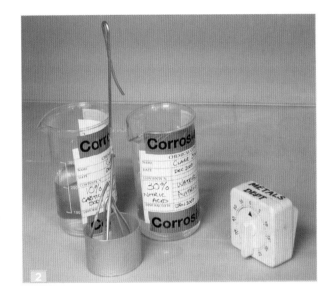

8. Immerse the aluminum in the nitric solution (Solution 2) for 1 minute. You will see this solution has worked as the smut is removed to reveal bright clean aluminum (photo 2).

9. Thoroughly rinse the aluminum under running, room-temperature tap water for 2 minutes (photo 3).

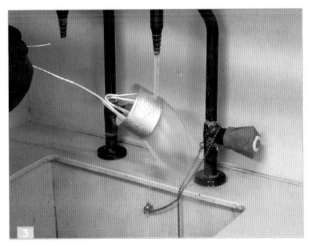

Anodizing the Aluminum

1. Fully secure the prepared aluminum in a jig. Hang the jig from the cathode, securing it well. As shown in photo 4, make sure there are even amounts of space around the aluminum when it is inside the sulfuric acid bath of the anodizing unit.

2. Depending on the size of your aluminum, slowly raise the voltage on the anodizing unit to between 4 and 16 volts. The voltage must be turned up slowly to prevent it from surging and cutting out. Increase the voltage until an almost invisible fine mist of bubbles appears in the bath. This mist means the anodizing process has begun (see photo 5). Leave the aluminum in the bath for 45 minutes. If you see more than a fine mist of bubbles, it may mean the voltage is too high. Shorter time in the bath may work to anodize the metal, but this depends on your setup.

3. Turn off the power to the anodizing unit. Remove the aluminum and thoroughly rinse it under running tap water for 2 minutes.

4. Optional: For maximum success, neutralize the anodized aluminum in the nitric solution (Solution 2, page 15) for 1 minute and then thoroughly rinse the aluminum under running tap water for 1 minute.

5. Remove the anodized aluminum from the jig and dry it thoroughly with a high-quality kitchen towel or paper towel. You are now ready to dye!

IMPORTANT

Your newly anodized surface is now porous and highly susceptible to contamination from grease and dirt. It is advantageous to dye the metal immediately. If this is not possible, the aluminum must be kept in an air-free environment until dyeing. A grip-seal bag with the air sucked out is perfect. Small pieces of anodized aluminum can also be kept in an airtight container with silica-gel moisture-removal sachets. If you have kept your anodized aluminum for awhile and it is not dying well, try dipping the metal in jeweler's pickle. This can help to revive the dyeing capacity.

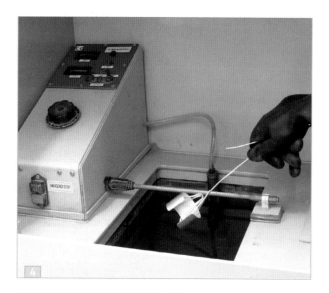

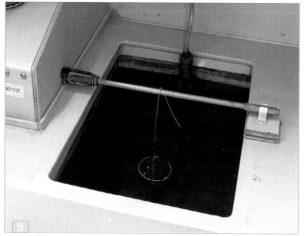

Flow Diagram of the Anodizing Process

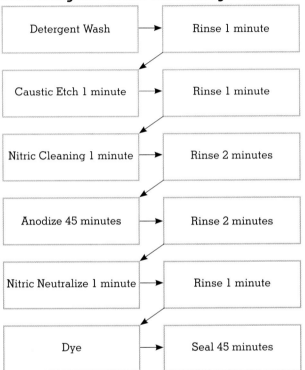

Detergent Wash	→ Rinse 1 minute
Caustic Etch 1 minute	→ Rinse 1 minute
Nitric Cleaning 1 minute	→ Rinse 2 minutes
Anodize 45 minutes	→ Rinse 2 minutes
Nitric Neutralize 1 minute	→ Rinse 1 minute
Dye	→ Seal 45 minutes

Sealing with Steam

After you have dyed your aluminum, you will need to seal the color. This can be achieved in a number of ways, but I prefer to steam my pieces.

To accomplish this, hang or stand the dyed aluminum pieces on end or diagonally in a steam container (see photo below). This positioning prevents pooling water and dyes from marking the metal. Steam the metal for no less than 45 minutes or for the same amount of time taken to anodize the metal. There is no real upper time limit for steaming dyed pieces but never allow the water to evaporate completely. This will cause the aluminum to start overheating and the dye will start to discolor and eventually disappear. You can easily test a cooled piece to see if it has sealed by dabbing it with a damp sponge. If the moisture is absorbed, the metal is still porous and needs to steam longer. If the moisture remains on the surface, it has sealed.

When steaming more than one piece at a time, marks may occur if the pieces are touching one another. Make sure to keep multiple pieces separated in the container. I use an old toast rack for this purpose, which works really well.

Dyed aluminum sheets being steamed to permanently seal the color

MATERIALS, TOOLS & EQUIPMENT

As with any craft process, preparing yourself with the right materials and tools is essential. In this section, I'll explain specialized equipment as well as how to use the supplies you already have. When purchasing your metal, always inform your supplier what you will be using it for, and their knowledge will point you to the correct alloy. Aluminum wire that is suitable for anodizing can be purchased from educational and sculpting suppliers. It is often sold as armature wire.

Shimpei Yamashita
Untitled, 2009
60 x 31 x 3.5 cm
Aluminum, copper, monofilament; soldered, formed, anodized
Photo by Helen Shirk

Aluminum

There are many different alloys of aluminum created to serve different purposes. Alloys are produced for added strength, increased malleability, and/or corrosion resistance. Some alloys, such as pure (99%) aluminum are more suitable for anodizing, but I have anodized many unknown alloys with great success, so it's always worth a go.

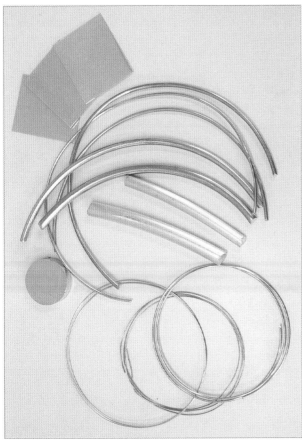

Raw aluminum can be purchased in many forms, including wire, bar, solid sections, and sheet.

Purchasing Pre-Anodized Sheet

As mentioned earlier, most makers use pre-anodized aluminum sheet, which is a great alternative to getting involved with the chemicals and special equipment needed for anodizing. You can purchase pre-anodized aluminum by the sheet or in packs. It's available in various finishes from matte to high polish, on one side or both. It is also offered in different thicknesses. The thickness of the sheet will be dictated by your design. In most cases, a thin sheet (up to 1 mm) is all you need. Aluminum sheets that are more than 1 mm thick can make fabrication difficult.

Bright, matte, and brush polished are just some of the finishes available for pre-anodized sheet.

Using an Anodizing Company

Another alternative is to source raw aluminum material and take it to an anodizing company. This is my preferred route. I give my size specifications to my aluminum suppliers who cut the sheets with a guillotine and then deliver them to my anodizing company. This method suits me well, giving me the freedom to choose whatever finish—from matte to mirror—and select whatever size I desire.

Finding a good anodizer is very useful, and essential if you are creating a piece that needs to be anodized after fabrication. I'm lucky to have a good company very close to my workshop that I have been using for over a decade and will tackle anything I give them.

Poppy Porter
Kaleidos Necklace, 2009
0.5 x 10 x 23 cm
Anodized aluminum, sterling silver
Photo by Angela Chan

Dyes & Inks

Dyeing anodized aluminum is great fun and very rewarding, but it can be very messy. Many of the substances used can dye other materials, surfaces, tools, and skin, so care must be taken. Avoid accidental dyeing by wearing rubber gloves and protecting porous surfaces with wipeable covers, such as oilcloth.

There are two main types of dye for aluminum—water-based dye and solvent-based dye. Industrial dyes specifically designed for anodizing are usually only available in large amounts, but some manufacturers are willing to sell small, sample quantities.

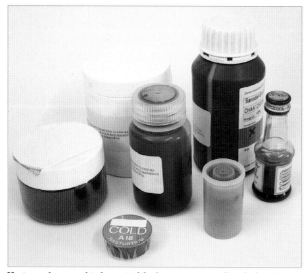

Various dyes and inks suitable for use on anodized aluminum

Solvent-Based Inks

Solvent-based inks, used mainly for industrial silk-screening, are also very effective for stamping and printing. They will act as a resist (or stop out) to a water-based dye, enabling vivid designs of opposing colors. These inks can be thinned with solvents for painting effects, and solvents are again required to clean brushes and equipment. Ink suppliers will sell their own thinners but will often advise of cheaper alternatives, such as white spirit, methylated spirits, acetone, and even good-old nail-polish remover. (Strangely, I find acetone-free polish remover works best!) When using these inks and thinners, make certain your workspace is very well ventilated.

Water-Based Inks

Water-based inks come in many forms. When sold industrially, they are usually in powder form to be simply mixed thoroughly with water. There are also water-based inks and pastes sold for commercial silk-screening, but these will not act as resists for additional colors. Industrial dyes can be costly, but they will achieve the best results for colorfastness, light-fastness, and variable degrees of intensity.

Water-based dyes perform better when warm. Creating some sort of warm-water or bain-marie bath for your containers will intensify the color. When using larger dye baths, however, this is not always

Hen Li
*Life of a Woman with Her
Hairpins II*, 2009
25 x 10 x 10 cm
Brass, aluminum, brass plates
with 18-karat gold; anodized,
pierced, screwed
Photo by artist

possible. Under this circumstance you can either make up new dyes using warm water, or cold dye the aluminum. The difference in intensity may be imperceptible and dyeing may take a little longer.

Household Dyes & Inks

You can dye anodized aluminum with many easily accessible commercial dyes, such as food coloring, fabric dye, and drawing inks. These dyes provide an easy way to get started coloring metal without investing too much in the process. Most are water-based products or powders to be mixed with water. Depending on the intensity of the color, these types of dyes are suitable for complete-coverage baths, for dip dyeing, and for painting. The effectiveness of these dyes varies from color to color. More often than not, mass-market dyes have less strength and light-fastness than industrial dyes, but they are extremely useful for beginners experimenting with the process. For some of my colors, I actually use a pigment designed to tint cake icing. It is all about being inventive and creative with your materials. If it's a dye, try it!

Pens

Certain types of pens provide a great way of decorating anodized aluminum. You can use overhead projector pens (OHP), compact disc markers, permanent markers, and certain types of graphic design markers to successfully create infinite patterns and designs. The use of a pen or pens is then followed by a water-based dye to create a background color. Some experimentation with the pens is required. Some pens will resist the water-based background dye, but some will mix with it and can run at the sealing stage. Later in the book, I'll talk about the behavior of all of these dyes in more detail.

Various pens suitable for use on anodized aluminum

Tools & Supplies

Along with the dyes, inks, and pens, a few other pieces of equipment are required.

Steamer Unit

The most important piece of equipment you'll need is some type of contained steamer unit to seal the final dyed colors. This can be anything from an old vegetable steamer to a tea urn. For my smaller pieces, I actually use a large saucepan with a metal steamer insert. I place the steamer in the saucepan and add water up to the base of the steamer. For my heat source, I use either a portable electric hotplate set on the lowest possible temperature or an electric food steamer, which has the

added benefit of a built-in timer. It is very important that the steam is contained and circulating rather than escaping.

Usually, all of these items can be cheaply sourced at flea markets, garage sales, and thrift stores. If you have the room, an old tea urn with working electrics and a lid is a perfect find. It allows for larger sheets to be steamed at a constant temperature.

In industrial settings, anodized aluminum is usually sealed by immersing the piece in boiling water or in a hot solution containing sealing salt. I have found that this method can strip out some of the color, so steaming is my preferred option.

Double electric hotplate with pans for warming dyes and steaming

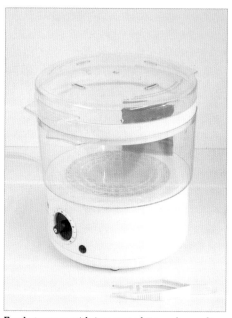

Food steamer with integrated timer for sealing

Additional Equipment for Dyeing

The following tools and supplies are useful, and many are necessary, for dyeing anodized aluminum. Everything on this list is easy to source, and you may already have most items on hand.

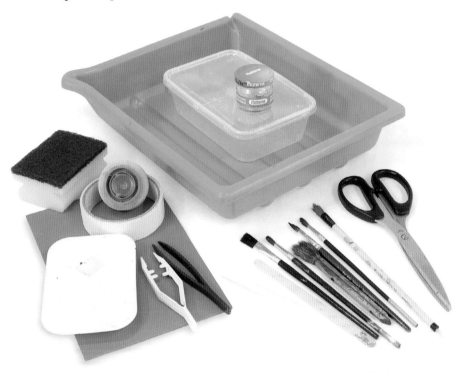

Clockwise from top: photographic chemical tray, takeout food container, jam jar, scissors, assorted brushes, craft sticks, toothpicks, craft foam sheets, plastic tweezers, plastic paint palette, masking and cellophane tape, mild kitchen scouring pad/sponge

DYEING KIT

Artist's paintbrushes, assorted sizes

Craft foam sheets

Decorative stamps

Garden trays or takeout trays (for dye baths)

General mark-making supplies

Glues, assorted types

Green kitchen scrub pads or pumice powder

Plastic containers

Plastic shelf liner with adhesive backing

Plastic tweezers

Rubber gloves

Scissors

Steaming unit

Tape, different varieties and widths

BASIC TOOL KIT

Doming block

Doming punches

Drill bits

Drill burrs

Emery papers, assorted grits

Files

Flexible shaft or hobby drill

Hammers, assorted

Jeweler's saw and saw blades

Mandrels

Pliers, flat-nose and round-nose

Rolling mill (optional)

Scribe

Snips

Stamps

Steel bench block or small anvil

Steel wire brush, for cleaning files

Tube cutter (optional)

Wire cutters

Left to right: pliers, wire cutters, tin snips, jeweler's saw, assorted files, steel bench block, assorted hammers

Left to right: steel wire brush, scribe, assorted mandrels, doming block and punches, emery papers (wet and dry), fine sanding block, assorted punches and stamps

Jewelry Making Tools & Supplies

Your jewelry-making experience and the tools you have at your disposal will influence the complexity of your final aluminum jewelry designs. For those with little or no jewelry making skills, I suggest using thin aluminum sheet (up to 0.8 mm). With thin sheet, you can cut out shapes with metal sheers or snips and finish edges with fine sandpaper. For those with more jewelry making experience, a basic tool kit will suffice (see page 24).

IMPORTANT

Aluminum can contaminate silver, which could be a problem when hallmarking your work. To prevent this, use separate tools for aluminum and for silver, and sweep your workbench and skin each time you change metals.

NOTE

The technical information within the step-by-step projects is weighted towards the aluminum process rather than basic jewelry fabrication. For more information on metalworking, specifically the forming and cold connection techniques used in the projects, I recommend *The Ultimate Jeweler's Guide* (Lark Crafts, 2010), *The Art & Craft of Making Jewelry* (Lark Crafts, 2010), and *Making Metal Jewelry* (Lark Crafts, 2006).

Tube cutter

SOLDERING KIT

Copper tongs

Cross-locking tweezers with wooden handle

Fire extinguisher

Flux

Flux brush or other applicator

Heat-resistant soldering surface (charcoal blocks, firebricks, or ceramic plates)

Pickle

Pickle warming pot

Safety glasses

Snips

Solder (hard, medium, and easy)

Solder pick

Soldering torch

Striker

Tweezers

Water for quenching

THE WORKING CHARACTERISTICS OF ANODIZED ALUMINUM

The fabrication and connection techniques used with anodized aluminum are slightly different from other metals more frequently used for jewelry. Because the melting point of aluminum is low—1220° F (660° C)—soldering and welding are difficult even for the most practiced hand. These applications also give off dangerous fumes. Color loss begins to occur at 752°F (400° C). This makes soldering components and findings to pieces impossible without damaging the color, but rather than a setback, this is an opportune time to get creative! The restrictions that anodized aluminum imposes are a catalyst for extra imagination. This element of challenge in construction makes for highly innovative pieces being produced.

Amy Hamai
Untitled, 2009
Collar: 16.5 x 16.5 x 2 cm
Aluminum, fine silver; riveted, formed, anodized
Photo by Helen Shirk

Storing & Handling

Anodized aluminum must be stored in a clean, dust-free, airtight container. If it is exposed to air for any length of time, the anodic film will seal itself, making it unreceptive to dye.

Always wear rubber gloves when handling anodized aluminum and avoid touching the main surfaces. Any direct contact will contaminate the tiny pores in the film with grease. Even the smallest of touches can lead to a relief fingerprint when dyed (see photo).

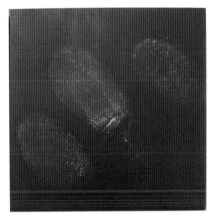

Fingerprints from handling anodized aluminum with ungloved hands

Cutting

Cutting dyed and sealed aluminum is generally not a problem. All the usual techniques can be applied, such as cutting with snips, sawing, piercing, using a guillotine, punching, and blanking. However, because of the softness of the core of the metal, you may want to use slightly larger saw blades to prevent stickiness and catching. The proper saw blade size also depends on the thickness of the metal. Along with beeswax, you can use chalk to ease the blade through the metal. Everyone has different methods for making tools work better, so experiment to find your own comfortable approaches.

Blanking

Blanking is a great technique for producing multiples of the same shape out of sheet metal. Steel sheets are pierced with the desired shape and used as a die. The metal to be cut is sandwiched between the inner and outer form. When pressure is applied, a scissor-like cut is made. This pressure can be achieved with a fly press, a vise, or sharp hammer blows to the inner part of the die. The RT Blanking system, a patented method, was invented to produce such dies, and specialized tools can be purchased to ease the process.

Pierced steel die (left) and shapes cut from it

Forming

Due to the brittle nature of the anodic film, bending, forming, hammering, and forging can cause a loss of color. Bending creates a crackled appearance that can be used for decorative effect, but with a careful hand, low pressure bends and forms can be successfully achieved without affecting color.

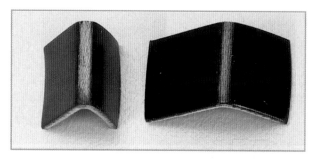

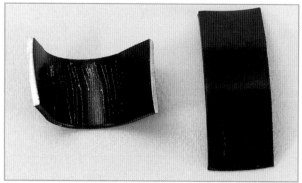

Examples of colored anodic film cracking/crazing under different degrees of pressure

Forging

Aluminum is fantastically malleable and lends itself perfectly to cold forging (see photo 1). The ability to hammer unique forms and twist and bend the metal opens up huge design possibilities. To extend the metal's workability, annealing may be required at regular intervals. All forging should be concluded before the anodizing process.

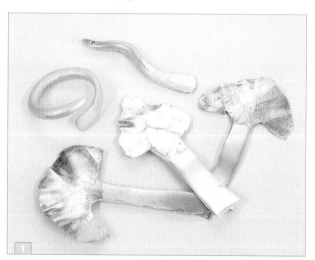

Far left: **Chung-Ting Yao**
Immost Feeling (Brooch), 2008
20 x 16 x 7 cm
Aluminum, copper, resin clay; anodized
Photo by artist

Left: **I-Ting Wang**
Blooming II, 2004
12.5 x 14 x 3.5 cm
Aluminum, sterling silver; dip dyed, photo
etched, hand painted, riveted
Photo by artist

Right: **Mandy Nash**
Brooch, 2009
10 x 6 x 0.5 cm
Aluminum, stainless steel; hand printed,
pierced, riveted, photo etched
Photo by artist

Annealing

It is possible to anneal anodized aluminum prior to
dyeing. Although usually unnecessary, annealing
the metal can help make it easier to bend. Use the
same annealing method for anodized aluminum as
you would for untreated aluminum.

Step by Step

1. Dot some liquid dishwashing soap in two
corners of the metal, avoiding the areas to be
decorated or dyed.

2. Place the aluminum on a clean, heat-resistant
surface. Heat the metal with a bushy torch flame,
keeping it constantly moving, until the soap turns
dark brown (photo 2).

3. Leave the metal on the heat resistant surface to cool.

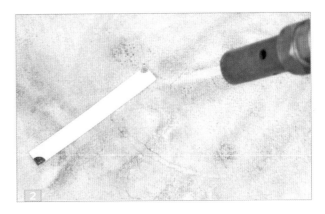

Tip

The soap acts as a visual indicator. When it
turns dark brown, you will know that the metal
has gained enough heat to anneal but not
enough to break the anodic film.

Left: **Clare Stiles**
Damask Cuff and Pendant, 2007
Aluminum; anodized, roller printed
Photo by artist

Right: **Tan-Chi Dandy Chao**
5ct Al-Diamond Ring, 2008
Ring: 3.5 x 2.5 x 1.5 cm
Aluminum, sterling silver; cut, dip dyed,
crown set
Photo by artist

Creating Texture

There are many ways you can achieve texture on anodized
aluminum and bring added dimension to your colored surface.
Experimentation is necessary to clarify how each process will
affect the color and how much pressure can be applied before it
is affected. All deep textures, such as hammer marks and stamping, should be made prior to anodizing. That said, many textures
can be successfully accomplished on colored sheet. Light hammering and roller printing are both very good ways to produce
subtle textures.

Dyed; roller printed
scrapbooking stencil

Tip

When hammering colored aluminum, if you apply more
pressure on the edges of the metal, you can achieve an
interesting color fade along a beveled edge.

Dyed; hammered with
ball peen hammer

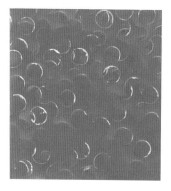

Dyed; hammered with
circle texturing hammer

Dyed; hammered with
cross peen hammer

Roller Printing

Relief patterns from wires, feathers, textiles, and threads can be transferred onto a metal surface with a rolling mill. Etched brass or steel plates can also be used to produce patterns and images, but as with other texture materials, it's necessary to experiment to ensure you apply the right pressure to achieve a good, clear result. It's important to know that when you roll metal through a press, any dyed images or patterns will become slightly distorted.

Dyed; roller printed wire

Dyed; roller printed leaf skeleton

Dyed; roller printed
sequin pop out sheet

Dyed; roller printed
plastic mesh garlic bag

Dyed; roller printed star sequins

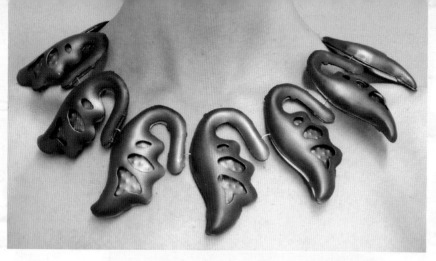

Rebecca A. Bandt
Pods, 2009
32 x 32 x 1 cm
Aluminum, copper, sterling
silver; die formed, textured,
anodized, pierced, oxidized
Photo by Helen Shirk

Die Forming

In the technique of die forming, sheet metal
is sandwiched between a wooden or
nylon die and rubber sheets.
The stacked sheets are
placed in a press or a vise
and pressure is applied. The
result is a positive relief of the
die shape. Die forming is a great
process to produce multiples of
the same form, difficult shapes,
or simply an individual form with
greater physical depth than a hammer alone can achieve. Die forming
is also an efficient alternative to the
lengthy process of repoussé. You can
press shallow forms in colored sheet, but
if deep shapes are desired, it's preferable to
press them before anodizing the aluminum.

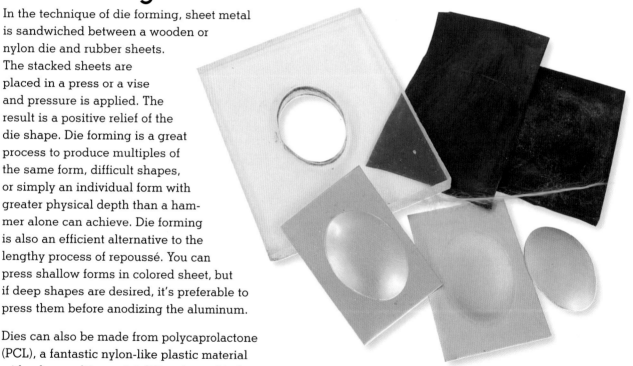

Clockwise from top: acrylic die, rubber sheets,
shallow die formed aluminum shapes

Dies can also be made from polycaprolactone
(PCL), a fantastic nylon-like plastic material
with a low melting point. When heated in hot
water, the PCL becomes extremely malleable
and can be formed in to any shape. Once
cool, it hardens into a hammer-proof and
press-proof material. PCL can be remolded
again and again, making it brilliant for tool
and form making.

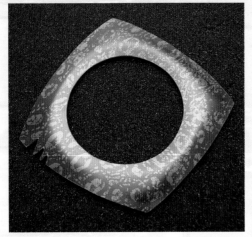

Right: **Mandy Nash**
Bangle, 2009
10 x 10 x 0.5 cm
Aluminum; hand printed,
pierced, pressed
Photo by artist

Far right: **Clare Stiles**
Anemone Bangle, 1996
12 x 4 cm
Aluminum, silver leaf;
hammered, reticulated,
anodized, brush blended,
riveted, gilt
Photo by artist

Reticulation

A careful hand is required for the reticulation of aluminum. Because the melting point of aluminum is so low, split seconds can mean the difference between a beautifully textured surface and complete meltdown! Again, experimentation is the key. With practice, you can achieve fantastic surfaces and edges and create gorgeous organic forms. All reticulation must be completed before the anodizing process. Reticulation can make aluminum very brittle and difficult to work with, so this process should be the last stage in your work before anodizing.

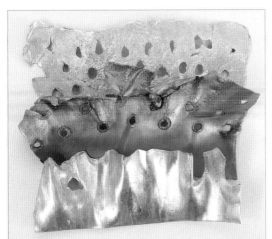

Reticulated test pieces of raw and anodized aluminum

Finishing

To finish and polish aluminum, use standard processes and tools such as files, emery papers, sanding blocks, and silicone or rubber burrs. The main difference is that aluminum will stick to the abrasive matter and cause clogging. To resolve this, keep a steel wire brush on hand for cleaning files, and use emery papers that can often be renewed. If you only have one set of files, remember to clean them thoroughly, and change any papers or burrs before working with silver again.

All aluminum surfaces must be finished to your required smoothness or texture before anodizing. The process will not hide flaws. After anodizing, even the smallest of marks or scratches

Clockwise from left: rubber burrs, split mandrel, rubber disk mandrel, disks cut from emery paper

Left: **Clare Stiles**
ring 1 |ri NG| noun
a circular band, typically of precious
metal and often set with gemstones,
a circular band of any material:
fried onion rings, 2009
2 x 4 cm
Aluminum, silver; anodized,
photocopy relief
Photo by artist

Far left: **Clare Stiles**
Superstylin, 2009
1.5 x 5 cm
Aluminum, powder coated steel;
anodized, photocopy relief
Photo by artist

on the metal will be visible. I prefer to use a split mandrel with the suitable paper inserted, chopping off the spent paper when necessary. For straighter edges, I use a rubber disk mandrel with emery disk attachments and replace them every now and again.

Depending on the depth of abrasion, use a file to remove harsh textures, and then move up through grades of emery papers or abrasive silicone wheels. If you are simply cleaning up a cut edge, a quick file and a pass with 400-grit emery paper will be enough to ensure a smooth, neat finish. Alternatively, you may wish to use a silicone abrasive wheel/ tip, but this option gets expensive as these attachments wear down fairly quickly.

A Bright Future

New technology for jewelry and metal work, such as digitally controlled laser cutting, milling and water cutting, rapid prototyping, and laser welding, are constantly evolving and are always interesting to learn about and consider. I recently had the opportunity to try out the digital engraver at the University of the Creative Arts in Farnham (UK). Head of metals Rebecca Skeels demonstrated how the engraver works and we did a test on a sealed anodized sheet (photo 3).

The hard anodic film called for a carbide cutting tool (the harder of the cutters) with a brass clamp to make a deep enough cut to get through to the color. The first two words were engraved at a depth of 0.2 mm, which left tiny chipping around the edges. We then tried a much shallower cut of 0.01 mm, which did not chip, and once repeated, gave a very clear smooth finish (photo 4).

The brilliance of this engraver is that any image, photograph, or word—however small—can be imported into the engraving program as a jpeg and cut into many different materials.

Right: **Lisa Vershbow**
Wrist Corsage, 2007
14 x 14 x 2 cm
Aluminum, acrylic, screws; anodized,
cut, folded, cold connected
Photo by Studio Munch

Far right: **Mandy Nash**
Necklace, 2009
20 x 12 x 0.5 cm
Aluminum, copper wire, stainless steel;
hand printed, pierced, riveted, stitched
Photo by artist

Findings & Attachments

The findings you use to construct anodized aluminum jewelry must be very strong to compensate for the lack of soldering. To create sturdy attachments, use heavy jump rings, hardened pin wires in silver and gold, and stainless steel wire. Incorporate the design of your findings into your piece, and use your wire to create wonderful findings that become an integral and effective part of the outcome.

Left to right: heavy jump rings, hard pin wire, head pins, split rings, beaded and hammered headpins

Rivets & Screws

Cold-connection techniques, such as riveting, are very important and useful for joining parts together. The connecting elements can be a visible part of the design or hidden from view. You can purchase small rivets (photo 5), screws, and micro nuts and bolts (photos 6) from model making suppliers, but it's simple and more creative and rewarding to make them yourself.

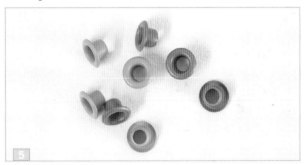

Geoff Palfreyman
Buttonhole Brooch, 1999
Closed, 4 x 10 x 0.8 cm
Aluminum, silver; riveted,
spring making
Photo by artist

Making a Simple Wire Rivet

When riveting aluminum, take care not to hammer the actual colored surface. Be gentle or use tape to protect the working area.

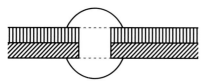

Cross-section view of a wire rivet

Step by Step

1. Determine where you wish to place the rivet, and use a center punch to dimple the aluminum at this point. Select a drill bit that is the same diameter as the rivet wire, and drill a hole through the dimpled aluminum.

2. Hold the rivet wire in a vise or pliers. Gently hammer the tip of the wire to form the beginning of the rivet head.

3. Thread the rivet wire through the pieces to be connected. Cut the wire slightly longer than the thickness of the pieces. (To determine the length needed, measure the thickness of the metal and add half the diameter of the rivet wire.)

4. Place the piece being riveted on a bench block with the existing rivet head facing down. Hammer the second head to secure the connection.

5. Lightly hammer both rivet heads until you are happy with their form.

Making a Flush Rivet

In a flush rivet, the rivet heads are level with the metal surface. After drilling the hole for the rivet, use a burr to create a recess around the hole. When hammered, the head of the rivet is countersunk into this expanded area.

Cross-section view of a countersunk
flush wire rivet

Making a Tube Rivet

Making a tube rivet is similar to making a wire rivet, except you'll need to first flare the ends of the tube (figure A) before hammering them over to secure the metal (figure B).

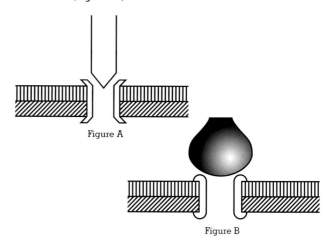

Figure A

Figure B

Tip

If you require movable parts, sandwich a thin piece of cardstock or cardboard between the metal layers before you start riveting. After the connection is made, soaking the paper in water makes it easy to remove.

Right: **Hen Li**
Candle in the Paper-Cut Flower, 2008
10.5 x 10.5 x 5 cm
Aluminum; pierced, riveted, anodized
Photo by artist

Far right: **Lindsey Mann**
Pendant, 2004
Each, 7 x 3.5 cm
Found plastics, aluminum, silver;
sublimation printed
Photo by Helen Gell

Making a Tube Rivet with a Spacer

Making a tube rivet with a spacer is an aesthetically pleasing and technically advantageous cold connection. This process requires two diameters of tubing. The smaller tube must fit perfectly inside the larger tube. Also, the larger tube, or "spacer," must have a more substantial, thicker wall than the smaller tube. This prevents sliding or sideways movement when the inner tube is riveted.

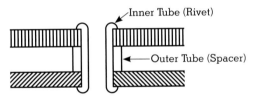

Cross section of a tube rivet with spacer

Step by Step

1. Determine where you wish to place the tube rivet, and use a center punch to dimple the aluminum at this point. Select a drill bit that is the same diameter as the inner tube, and drill a hole through the dimpled aluminum.

2. Cut both tubes to size, considering the space required between the layers and the thickness of these layers. The inner tube should slightly protrude when all the layers are constructed.

3. Anneal the inner tube. Using a center punch or scribe, flare one end of the tube.

4. Thread the flared inner tube through the first drilled hole, then through the spacer tube, and finally through the second drilled hole. Flare the second end of the inner tube to secure the metal layers.

5. Place a dapping punch in a vise, then use a hammer start to curl the tube outwards. Hammer the tube evenly on both sides to secure the rivet and planish the surface.

Tip

When hammering, if you find the layers sliding sideways, it may be helpful to make yourself a little tool and slide it between the layers to hold them apart (figure C). Cut a small piece of hardwood or nylon sheet to the same width as the gap. Cut a slit on one end of the material that is the width of the outer tube. Slide the material between the layers you are riveting. This will allow you to hammer without "squashing" the piece.

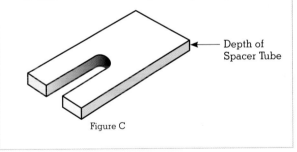

Figure C

Adhesives

Although I'm not a big fan of using glue in jewelry, sometimes it is required. If so, use a two-part epoxy. This type of adhesive has a degree of elasticity when dry, as opposed to brittle, fast-drying glues that can snap with just a small amount of pressure.

DESIGN BASICS

Jon M. Ryan
Untitled, 2009
8.6 x 3.2 x 1.8 cm
Aluminum, sterling silver; carved, fabricated, anodized
Photo by artist

When designing jewelry, take the time to thoughtfully consider color, inspiration, and scale. Developing and applying these practices will contribute to the overall integrity of your work.

Color

Choosing colors for your jewelry may come about in several ways—your design or inspiration may dictate the colors you wish to use; you may be limited by the dyes you have available; or your colors may simply evolve through experimentation. Whichever path you take, it may be helpful to think about some basic color theory to aid your choices.

Looking at the color wheel on page 39, you can see many ways of mixing and blending color. All of these theories can be applied to dyeing anodized aluminum. Using the three primary colors of red, blue, and yellow, an infinite spectrum of possibilities can be achieved.

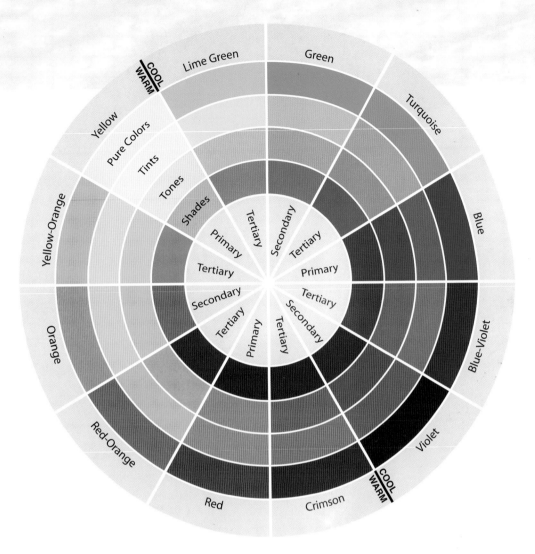

Color Glossary

Hue	The pure color
Tint	Pure hue plus white (or water) to lighten the color
Complement tint	Tint plus a small amount of grey or opposite color to tone down the color
Shade	Pure hue plus black to darken the color
Tone	Pure hue plus grey

Color Combination Glossary

Monochromatic	Use of the same hue with different tones or tints
Complementary	Opposite colors on the color wheel
Harmonious/ Analogous	Colors next to each other on the color wheel
Triadic	Colors evenly spaced around the color wheel
Neutral	White, black, grey, and beige

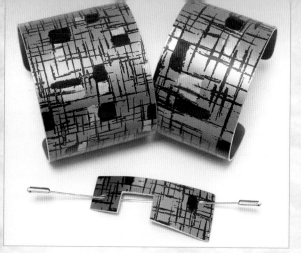

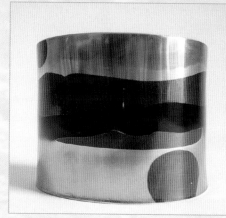

Far left: **Clare Stiles**
Crosshatch Cuffs and Brooch, 2007
Cuffs, 5 x 6 cm
Brooch, 1.5 x 7 x 2.5 cm
Aluminum; anodized, silkscreened,
hand painted, submersion dyed
Photo by artist

Left: **Michael Peckitt**
Common People Collection, 2009
5 cm wide
Aluminum; anodized
Photo by artist

Inspiration

> 1. *Something or somebody that stimulates the human mind to creative thought or to the making of art*

Inspiration for design comes from countless forms and sources. Nature provides us with an incredible supply of color and form. From the leaves on the trees, the fish in the sea, and the birds in the sky, nature will never fail to inspire creativity. Photography, art, and architecture provide obvious stimulation, but everyday events, memories, and objects can also influence our designs. A trip to the market, a cup on a table, a poem, map, or tool could all trigger thoughts and inspirations for our work.

When viewing the world with open eyes and a creative mind, inspiration really can come from anywhere! A perfect example of this is my studio mate Liz, who has a fascination with cogs and rusty washers. For myself, all of the above apply. I love to observe extraordinarily beautiful things, manmade or natural, and reflect on them in my work. I am heavily influenced by graphics, furniture, textile, and architectural design. My memories and photographs of childhood and travels all play a huge part in inspiring me.

Scale

Another important consideration when designing jewelry is the scale of the pattern or image. For instance, if you are designing a pair of small earrings, the pattern needs to be small for the effect to be captured when the metal is cut to size. You may want to be able to see a repeat of the pattern or use just an enlarged section. Either way, pattern scale must be considered in the design.

Tip

Whether or not you are interested in following fashion, it should not be ignored as a source of color inspiration. Fashion-forecasting websites have fantastic mood boards and color charts, which make a great starting place for designing with color.

DYEING

Of all of the coloring and decorative techniques used with anodized aluminum, submersion dyeing may be the most important. For this process, you will mix a water-based dye with warm or cold water in a suitable container. The container must be large enough to hold the aluminum, which needs to be completely immersed to achieve complete coverage of the chosen color.

Tip

For all of the dyeing techniques, it's best to work on a larger piece of metal than needed so that any marks acquired during the sealing process can be cut away from the edges.

Kate Fehr
Firefly, 2009
Necklace, 30 x 17 cm; each earring, 6 x 4 cm
Aluminum, silver; anodized, soldered
Photo by David Fenton

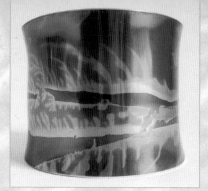

Far left: **Michael Peckitt**
Stormy Weather Bangle, 2009
5 cm wide
Aluminum; anodized
Photo by artist

Left: **I-Ting Wang**
Blooming I, 2004
11.5 x 13 x 3.5 cm
Aluminum, sterling silver; dip dyed,
photo etched, hand painted, riveted
Photo by artist

Sequence

When using any of the decorative dyeing processes, you'll usually start, or more often finish, with full submersion into a color of your choice. Before you begin a project, it's important to decide if you want to do your decorative work on top of the chosen background color or add the background color after completing the decorative work. When you work on top of an existent background color, you will be blending any colors you add. When you choose to create a pattern first, it will act as a relief and the submersion will fill in the gaps with color. The examples below show the importance of the application order of dyes and resists. With just two colors, very different effects are achieved.

Cellophane tape applied in strips; dyed with bronze water-based dye; tape removed

Dyed with bronze water-based dye; cellophane tape applied in strips; dyed with turquoise water-based dye; tape removed

Dyed with turquoise water-based dye; cellophane tape applied in strips; dyed with bronze water-based dye; tape removed

Cellophane tape applied in strips; dyed with turquoise water-based dye; tape removed; dyed with bronze water-based dye

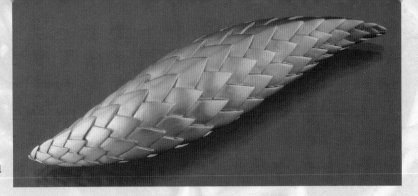

Jon M. Ryan
Untitled, 2009
7.7 x 1.9 x 1.3 cm
Aluminum, sterling silver;
carved, fabricated, anodized
Photo by artist

Timing

The timing of a submersion bath depends on the strength of the dye you have mixed and the color you desire. Strongly mixed dyes take just a few seconds to color anodized aluminum. Weaker mixes can take up to a few minutes. It is important to note that if you choose to over-dye a new color, leaving the aluminum in the dye for a longer amount of time can weaken the effect.

Mixing & Over-Dyeing

It's possible to mix colors through submersion. To accomplish this, you dye the aluminum one color and then submerge it into a second color. This over-dyeing creates a new color that is a mixture of the two. This process works because the secondary color slightly washes out the pores that contain the original color and replaces them with itself. You may notice that if you continue to over-dye with the same color baths, the secondary dye will slowly become contaminated with the first color.

Some dye colors are much stronger than others, and this can have an impact on over-dyeing. Reversing the order in which the dyes are applied is a good test to see if and how the final color is affected. For instance, if I use a reasonable-strength yellow as the second color in an over-dye, the yellow will overpower most other colors. I achieve vastly different oranges depending upon whether I use red or yellow as the first color (see photos below). As you develop your practice, you may choose to mix favorite colors into a single bath. This makes over-dyeing unnecessary and helps you to attain consistency.

Dyed yellow sheet dipped in red dye

Dyed red sheet dipped in yellow dye

Selective Over-Dyeing, Gradation & Blending

You can use the over-dyeing process on selective parts of a piece. Rather than completely submersing the aluminum, try dipping just one end in the bath. You can choose to create two distinct, side-by-side colors, or you can create a blended effect by gradually removing the piece from the dye a little at a time.

This blended effect will also change depending on whether the aluminum is wet or dry. By carefully drying the metal with a kitchen towel between baths and timing its removal from the second bath just right, the blending can be flawless. A brush is a helpful tool for blending and carefully gradating colors. Use a brush to mix water into the dyes to gain different strengths of tone. Over-dyeing wet aluminum produces a completely random, almost tie-dyed effect where the dye runs in channels between colors.

Gradual removal from the dye bath produces gradation in color

Dipping to a single depth creates a definite edge between colors

Brush blending achieves a flawless blend

A wet surface causes the dye to run up the sheet in channels

Inks

As seen in these samples, solvent-based inks are best for creating contrast through techniques such as dripping, sponging, spraying, flicking, and of course, painting. Experimentation with a variety of tools can yield infinite, beautiful designs. Get in the habit of keeping notes as you would for a recipe (see photo). Record the order in which you applied dyes, which tool you used to achieve a particular pattern, etc. Writing down all the little details will make repeating your designs much, much easier.

Brushstrokes of thinned green ink

Brushstrokes of thinned red and yellow inks; over dyed with dark brown water-based dye

Yellow ink rolled onto acetate, cut away with a pencil, and monoprinted; over dyed with purple water-based dye

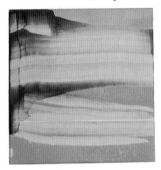

Dark purple ink dragged over sheet with part of old credit card; over dyed with turquoise and bronze water-based dye

Thinned inks brush painted; over dyed with bronze water-based dye

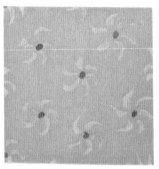

Yellow and orange inks applied by brush; over dyed with pale teal water-based dye

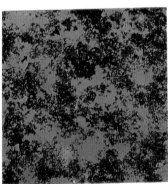

Turquoise ink applied with sponge; over dyed with red water-based dye

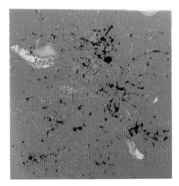

Yellow and brown inks flicked off of stiff brush; over dyed with turquoise water-based dye

Monoprinted leaf with red ink; over dyed with green water-based dye

Dyeing   45

Resists

By using resist materials, you can achieve contrasting colors, depth, and the illusion of texture in your work. There are many ways to create complete or partial resists on anodized aluminum.

PVA glue applied and dried; dyed in pink water-based dye; glue peeled away

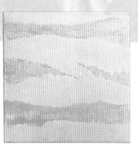

Dyed in bronze water-based dye; ripped masking tape applied ; dyed in yellow water-based dye; masking tape removed

PVA glue piped on as outline and dried; thinned inks painted within the gaps; steamed; glue peeled away and inks removed with thinner

Dyed in turquoise water-based dye; masking tape applied; dyed in pale purple water-based dye; tape repositioned; dyed in pale purple water-based dye; tape removed

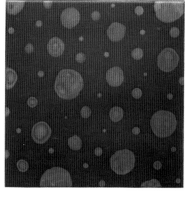

Dyed in purple water-based dye; PVA glue dots applied and dried; dyed in turquoise water-based dye; glue dots peeled away

Dyed in yellow water-based dye; PVA glue applied with edge of old credit card section and dried; dyed in purple water-based dye; glue removed

Dyed in turquoise water-based dye; cellophane tape applied; dyed in bronze water-based dye; cellophane tape removed

Dyed in red water-based dye; PVA glue applied, pulled across sheet with toothpick, and dried; glue removed

Dyed in turquoise water-based dye; cellophane tape applied; dyed in yellow water-based dye; cellophane tape peeled away

Pens

Whether you are drawing a purposeful picture or creating a random pattern, using certain pens can be a great and simple way to add color and pattern to your work. Different pens create different results. I encourage you to play with your pens and test their mixing ability, their ability to act as a resist before and after dyeing, and their colorfastness during the sealing process.

Turquoise and pale turquoise graphic design marker; over dyed with water-based pink dye

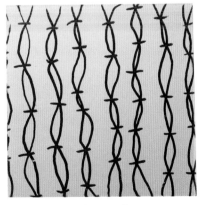

Moss green graphic design marker

Pink and yellow swirls and red dots drawn with permanent overhead projector pens; over dyed with water-based purple dye

Black, purple, and green graphic design markers; over dyed with water-based grey dye

Black compact disc marker; over dyed with water-based yellow dye

Pale ochre graphic design marker; over dyed with water-based olive dye then water-based red dye

Tapes & Glues

Tapes make excellent resist materials. When masking tape is placed onto aluminum, it will prevent the surface it covers from being dyed. Various effects can be achieved through different applications. Experiment with several applications and learn which ones you prefer. For example, you can cut the tape into shape, or tear it into random pieces. You can lightly place the tape on the metal, rub it down thoroughly, or create wrinkles in it. Try dyeing the aluminum before and after applying the tape. Between submersion baths, try removing and replacing the tape in different positions. As with most tapes, masking tape will leave behind a sticky residue. This will have an effect on any dyeing done after the tape is removed, but it can also make for interesting outcomes, so use it to your advantage! Removable glues and solvent-based dyes also resist water-based dyes and allow for contrasting color combinations.

Additional Techniques

Other methods for resisting a dye, such as photocopy transfer, acetate, and PNP relief, are slightly more involved than the simple application of tape or glue, but can produce more defined patterns or pictures. These methods are great for producing multiple pieces using the same patterns.

Making & Transferring a Photocopy Resist

Photocopy resist is a fantastic process for achieving graphic and photographic type images with contrasting colors. Practice and play with the process to find your own individual style.

Step by Step

1. Make a photocopy of a pattern or image you would like to reproduce on your work (photo 1). Reverse the image, especially if it includes text. Important: Laser copiers will not work for this process. Their inks are unsuitable.

2. Position the photocopy face down on an anodized aluminum sheet, and secure it with a little tape at the edges (photo 2).

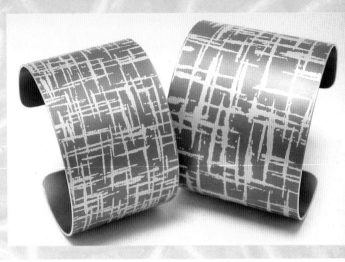

Right: **Deepti Kumar**
Marigold, Earrings, 2007
Each, 3.6 x 4.5 x 0.4 cm
Aluminum, silver; hand
painted, photoetched, riveted
Photo by Sussie Ahlburg

Far right: **Clare Stiles**
Crosshatch Cuffs, 2007
Each, 5 x 6 cm
Aluminum; anodized,
silkscreened, hand painted,
submersion dyed
Photo by artist

3. Place a very small amount of acetone or other such solvent on a cotton ball or kitchen towel. Gently dab the back of the photocopy, especially the area over the image. Continue to rub firmly but gently until the acetone starts to dry. Carefully lift a corner of the paper to see if the ink has transferred to the metal, then lift the paper to reveal the full image (photo 3).

Troubleshooting: Using too much acetone can cause an image to smudge and can block the pores of the anodic film, preventing them from accepting a new dye. If this happens, use less acetone, or let the cotton ball dry slightly before applying it to the photocopy. You may have to have a few tries to get it just right.

4. Dye the sheet and steam it to seal the color. Remove the photocopy ink with thinner to reveal the aluminum-colored image.

Over-Dyeing with a Photocopy Resist

Although you can apply a photocopy transfer before dyeing an anodized aluminum sheet, the best application of this technique is over a dyed surface.

Step by Step

1. Follow the process described on pages 48 and 49 to transfer a photocopied image onto a dyed aluminum sheet.

2. Submerge the sheet into jewelers' pickle or a weak solution of nitric acid (photo 4). This bath will remove all color except that which is under the transferred image (photo 5). Thoroughly rinse the sheet.

3. Over-dye the aluminum with a second, contrasting color and seal the sheet (photo 6). Note: This

method works best when the second dye choice is darker than the first.

4. When the sealing is complete, use acetone on a cotton ball to remove the black photocopy ink (photo 7). This reveals the image in your first dye color.

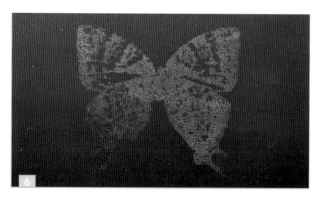

Making & Transferring an Acetate Resist

This technique produces the same results as the paper photocopy and acetone method. It may take a little practice to get right. Too much heat from the iron can seal the anodic film, but too little heat can yield an incomplete image.

Step by Step

1. Select an image and print it onto photocopy-compatible acetate (photo 8).

2. Set an iron to a low heat. Position the acetate, ink side down, on the anodized aluminum sheet (photo 9). Place a piece of paper or cloth over the acetate.

3. Use the iron to transfer the image from the acetate to the aluminum. Press the iron down firmly and keep it moving at all times for a couple of minutes (photo 10). Leave the piece to cool. Important: The metal will be hot, so take care not to touch it too soon.

4. Peel the cooled acetate off the aluminum sheet. The photocopy ink will be left on the metal surface (photo 11).

continued on next page

continued from previous page

5. Over-dye the sheet (in this sample, purple dye was used), and steam it to seal the color (photo 12).

6. Remove the ink with a thinner to reveal the original color, in this instance red (photo 13).

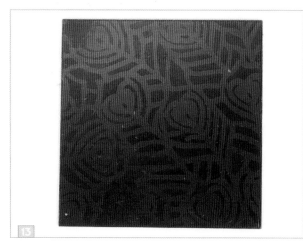

Press & Peel (PNP) Paper Resist

The same process used for making an acetate resist can be used with press-and-peel paper (also known as PNP) with one exception—PNP paper can be printed by a photocopier or by a laser printer. Use pressure and heat from a warm iron, or heat press to transfer the printed image onto the metal. Allow the materials to cool completely before peeling off the paper. The aluminum should be sealed before the ink is removed with a solvent.

PNP paper was ironed onto dyed aluminum, the paper was peeled off, and the resist remained in place. The aluminum sheet was over-dyed and steamed (top photo), then the ink was removed to expose the original dye color (bottom photo).

PRINTING

There are many possibilities for printing on anodized aluminum. Traditional printmaking and textile-printing techniques can both be applied. The tools used range from very basic to complex. The inks used are specifically designed for printing, usually silk-screening on an industrial scale, and can be solvent- or water-based. Printing on aluminum requires the slower-drying, thicker consistency that these types of inks provide.

If you plan to use contrasting colors, it's important to use solvent-based inks. They will "stop out" or resist any submersion dyeing you may choose for a base or background color. If you choose to cover a whole surface with inks, to leave the background the natural aluminum color, or want to blend the colors, using solvent-based inks is less important or unnecessary.

Lorraine Gibby
Corsages, 2009
Smallest: 2 x 2 x 1 cm
Aluminum, white precious metal, semiprecious stones; printed, dyed, textured, riveted
Photo by Steve Speller

Stamping

Stamping is used to repeat images or patterns on anodized aluminum. Stamps can be bought from most craft suppliers, but handmade ones can be cut from linoleum, craft foam, and nylon sheets. To make small, handmade stamps easier to handle, glue them onto a small block of wood or a similar rigid surface. Marking the wood block with an arrow helps you position the stamp when printing. Pens can be used as well as standard printing ink to add the color to the stamp, but you'll need to work fast because pens dry quickly!

Examples of purchased and home-made stamps

Yellow ink on purchased stamp block; over dyed with olive then purple water-based dye

Teal design marker on purchased stamp block; over dyed with red water-based dye

Orange ink on craft foam stamp; over dyed with purple water-based dye

Olive green graphic design marker on craft foam stamp; over dyed with red water-based dye

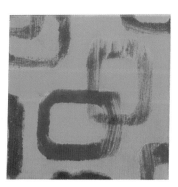

Teal graphic design marker on craft foam stamp; over dyed with turquoise water-based dye

Pale aqua graphic design marker on craft foam stamp; over dyed with bronze water-based dye

Monoprinting

Monoprinting essentially uses inks on one surface to print onto another, producing a pattern or painting. This technique creates one-of-kind images with a uniqueness and spontaneity that is very different from direct painting.

Thick acetate cut to size is an ideal surface for applying ink, but cheaper alternatives can be used, such as any semi-flexible, wipe-clean surface you may find around the house. A print roller is also needed. This tool can be purchased at most art supply stores.

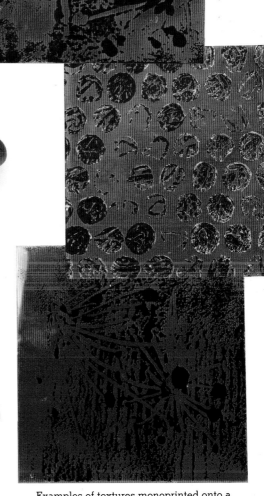

Monoprinting equipment, clockwise from top: black printing ink, print roller, brush, toothpick, craft stick, corrugated card, bubble wrap, acetate sheet

Examples of textures monoprinted onto a dyed sheet

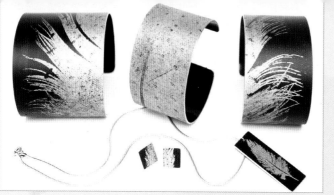

Monoprinting

Step by Step

1. Apply a thin, even layer of ink onto the acetate with a print roller. (This provides a base layer of ink for you to work into.)

> The base may be one color, random flicks of color, or a row of colors that blend as they meet.

> You can layer textures onto your image, using inked materials such as cardstock or popped bubble wrap, but take care to keep a similar surface level across the acetate plate. This prevents unwanted blank spots when the plate is printed.

2. Work into the inked plate to create your image, using any number of tools, brushes, found objects, and texture materials (photo 1). You need to work fairly quickly so the inks won't dry.

> There are infinite possibilities for creating an image in the ink and generating marks. For example, drag a wooden craft stick on its side to create a pattern, use the end of a pencil to draw a more intricate and purposeful image, or flick a decorating brush across the surface to add texture.

3. Carefully lower the aluminum—anodized side down if only one-sided—onto the image on the plate. Apply even pressure to the back of the sheet with a clean roller, heavy books, or just your hands. Take care not to slide the surfaces, or you will end up with a great big smudge!

4. Turn the plate over and very carefully peel off the acetate to reveal your image (photo 2). Leave the aluminum to dry for about 1 hour.

5. Seal the printed aluminum with steam, and clean off any remaining ink with a solvent or thinner. Clean the acetate sheet and any tools while the ink is still wet. Use water or solvents, depending on the ink you have used.

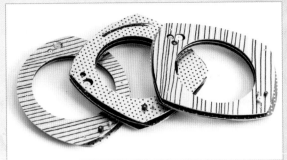

Linoleum Printing

To create a linoleum print, or lino print, you cut linoleum with a tool that removes the surface bit by bit to produce a picture or design. The blade on a linoleum cutter is sharp, so watch your fingers and handle with care. As shown in photo 3, the final image is printed in reverse, so keep this in mind when you are designing your image.

3

Step by Step

1. Use a pencil to draw your design onto a linoleum sheet.

2. Before cutting the linoleum, take a moment to verify which elements of the image are to be removed. Which are the positive (to be inked) sections and which are the negative (to be cut away) sections?

3. Carefully follow the drawing with a linoleum cutter. Don't cut your image too deep, just enough to produce the negative space (photo 4).

continued on next page

4

continued from previous page

4. Ink the linoleum with a roller (photos 5 and 6), and print the image onto the aluminum. At this point there are several options: you can stop here and over-dye the aluminum in a submersion bath; you can clean the linoleum, re-ink it with another color, and print the aluminum again, or you can create a new layer with an altogether different image.

Silkscreening

Silkscreening is the most complex but also the most rewarding printing method for anodized aluminum. Intricate detail can be achieved again and again from the same screen.

The process involves drawing inks over a screen to print the relief on the surface below. Producing the highest-quality silkscreen requires specialized equipment and materials. Simple silkscreening kits, however, can be bought from art supply stores without investing vast amounts of money. If you wish to learn how to make very precise or detailed designs, I suggest contacting the textile or fine-arts department at your local college. They may offer classes or be willing to help.

A silkscreen is a frame, normally made of wood or metal, with a fine nylon mesh stretched over it. The mesh is measured in Ts (threads per inch) or LPIs (lines per inch). The finer the mesh, the higher its LPI or T number. Designs with a high amount of detail require a finer mesh.

An exposure unit has a glass bed onto which a silkscreen and an image are laid. Its lid is brought down, and a vacuum pulls the screen and image taut to the bed. UV lights are then shone through the image to harden the emulsion.

Before you start designing an image, it's helpful to know that silkscreening can only produce lights and darks, no middle tones. If you desire grays in your design, these need to be pixelated (made into tiny dots, as in newspaper printing) or converted to lights and darks, which can easily be achieved using the contrast adjustments on your computer.

An acetate image
and a finished silkscreen

Step by Step

1. Photocopy your desired pattern or picture onto photocopy acetate. The black on this acetate will eventually be the negative on the screen and the area that will print. What you see as a print on the acetate is what you will print on the aluminum.

2. Blacken the print on the acetate with black ready-mix tempera or poster paint. This improves the opacity of the image and prevents UV light from seeping through it. Using a handful of paper towels, gently rub the black paint over the copied image. The paint will stick to the toner and can be wiped off the clear acetate.

3. Evenly coat the mesh screen with a light-sensitive film or emulsion under safe red light (non UV, such as those used for developing photographs), and let it dry.

4. Place the screen on top of the acetate and into an exposure unit. UV lights will shine up through the design to harden the coating in parts. Exposure times vary, depending on the mesh you're using and the quality of your photocopy acetate.

5. Use a gentle jet of water to wash away the emulsion that hasn't been hardened, leaving a stencil of the design to be printed.

6. Place the anodized aluminum under the screen and line up the image above the metal surface. Important: The screen should not move during this process, so either enlist another pair of hands to hold the screen down or securely clamp it to your work surface. (Advanced screenprinters use a hinged clamping frame, but the above options are fine.)

7. Run ink along one end of the screen and steadily pull it across with a squeegee (photo 7). (This process, which is called flooding, forces the ink into the fine holes of the mesh.) Pull the squeegee across the screen a second time to complete the print.

8. Carefully lift the screen from one end (photo 8) and remove the aluminum to reveal your printed sheet. Let the ink dry completely. (You can speed this up with a hair dryer.)

9. Submerge the screenprinted aluminum into a dye of your choice. Rinse off the dye with water and thoroughly pat the metal dry to avoid marks (photo 9). Keep the metal clean and safe until you seal it. After sealing, remove the ink with a thinner.

10. While the ink on the screen is still wet, thoroughly clean it off with a solvent, newsprint, and an absorbent towel. Regularly replace the newsprint and towel until the screen is absolutely clean with no color residue. This is a very smelly process, so make sure the area is well ventilated.

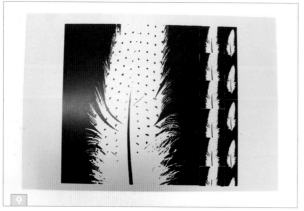

Sublimation Printing

Sublimation printing is a common process used for full-color printing on mugs, T-shirts, mouse pads, and other commercial items. Sublimation inks are used in inkjet printers and printed onto transfer paper. When high temperatures and pressures are applied, the inks turn into a gas and transfer onto the object. The results are extremely hard wearing.

When anodized aluminum is sublimation printed, wonderful full-color images are produced, but the process also has its downsides. It's expensive to set up, the colors don't always transfer exactly the same, and there is a very slight fuzziness of detail. This method requires specialized equipment: a flatbed heat press, sublimation inks, and an inkjet printer specifically devoted to these inks. If you're interested in experimenting with this method but don't want to make the investment, visit your local mug and T-shirt printer, and ask if they can help. They may not know about anodized aluminum and may be as interested to see the results as you are!

Digital Printing

Digital printing is a relatively new process in which images are applied directly onto the aluminum. This method produces fantastic full-color photographic images and opens up a whole new world for anodized aluminum. The images are printed from a special flatbed printer using suitable inks. Only a few companies are providing this service, and unfortunately, it is very expensive.

TIPS & TROUBLESHOOTING

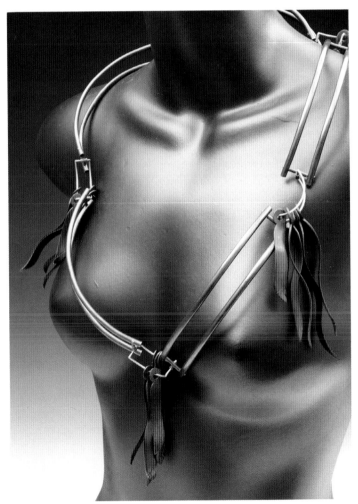

Devon Hoffman
Untitled, 2009
39 x 25 x 2.5 cm
Aluminum, copper, nickel; riveted
Photo by Helen Shirk

The Anodizing Process

If your metal doesn't accept dye after it has been anodized, this may be due to a number of reasons. You may have an idea of what has gone wrong, but if not, work through these questions:

- Did you jig your piece securely enough?

- Did the power overload and cut out during the process?

- Did you set the power too low?

- Did you follow the cleaning stages correctly?

- Does the connection to the anode need cleaning?

If your jig burned away during the anodizing process, the power is too high. Be sure to use a suitably sized jig wire to prevent this.

Water marks or patches in the color are very likely to occur when the electrolytic solution has not been thoroughly rinsed off the aluminum after anodizing. The metal should also be dried thoroughly, either by hand or by a warm air source—otherwise, pooling water can cause a mark.

Sally Lees
Stripe Cufflinks, 2007
Square, 1.9 x 1.9 cm
Round, 1.6 x 1.6 cm
Aluminum, silver; hand
painted, dyed, riveted
Photo by Full Focus

The Dyeing Process

Always store anodized aluminum sheet in an airtight bag or container. Silica sachets are useful in this situation, as they will absorb moisture away from the metal.

Have a look in your local craft shop or on websites for stencils, stamps, permanent inkpads, and similar supplies to use as a source of inspiration and for application techniques.

If you're a beginner, experiment with fountain-pen ink, food coloring, and clothing dyes before committing to the expense of specialty dyes.

When you've used a water-based dye, some of the color will escape during the sealing process and pool as condensation. This can be used decoratively, but to prevent it, follow these tips:

- Hang the metal or lean it on a corner so the droplets of condensation can run off easily. (Water marks will occur if the work is steamed flat.)

- Move the metal occasionally during steaming.

- Do not lean metal pieces against each other, as colors will run from one piece to another.

- Always work on a larger piece of metal than needed so you can remove any marks.

Use a gentle scouring pad or fine pumice to gently rub off any scum that forms on the metal surface during sealing.

When you handle a piece after sealing, if your fingers get colored or you make fingerprints in the dyed surface, it means the metal needs to be sealed longer. As long as the water supply doesn't run out, you cannot damage a piece by sealing it longer.

Not all pens are colorfast during the steaming process. Always do a test run before committing to a piece of work.

Most dye colors can be stripped from an unsealed aluminum surface with jeweler's pickle or diluted nitric acid, such as that mixed for cleaning.

If you've steamed a piece of aluminum but are unhappy with the result, you can remove the hard anodic film by submersing the metal in a solution of caustic soda and water. The time this takes to work depends on the strength of your solution and the thickness of the anodic film. You'll see the dye coming off almost immediately, but the film goes deeper than the colored layer. You must leave the piece in the solution longer to reach the core of the aluminum.

The Finishing Process

When filing edges, file away from the metal rather than toward it. This will help reduce small chips and will result in a nice clean line.

When filing, keep a steel wire brush handy and use it regularly to clean clogged aluminum from the teeth of the file.

The Projects

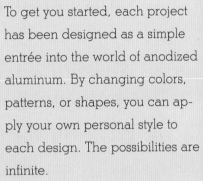

To get you started, each project has been designed as a simple entrée into the world of anodized aluminum. By changing colors, patterns, or shapes, you can apply your own personal style to each design. The possibilities are infinite.

Whether solvent-based or water-based, all of the dye colors used for the projects can be mixed from a palette of red, blue, yellow, black, and turquoise. Clear ink is also useful for lightening the hue of a color.

For each project, you'll need a Dyeing Kit, which can be found on page 23, and a Basic Tool Kit (page 24). For a few projects, you'll need to anneal or solder metal. The tools and materials for these techniques are listed on page 25. All additional supplies are listed before each project.

Charm Bracelet

A fantastic project for the beginner and the practiced hand alike, this bracelet requires only a small amount of aluminum and is so flexible—add beads, mix colors, create a new pattern, use leftovers from projects past—go as far as you like! I used two colors and kept the design small and simple with a plain chain and clasp, but you can adapt these variables to your unique taste.

YOU WILL NEED

Anodized aluminum sheet, 1 mm, 10 x 10 cm

Sterling silver round wire, 0.8 mm, approximately 100 cm

Sterling silver round wire, 2 mm, 1 cm long

11 heavy silver jump rings, 5 mm

10 beads of your choice, 5 mm

10 beads of your choice, 3 mm

10 silver headpins

Dyeing kit, page 23

Basic tool kit, page 24

Soldering kit, page 25

Permanent marking pen (project features olive green)

Water-based dye (project features pink)

Small oval mandrel, approximately 1 cm long

Disk cutter or jeweler's saw

Rolling mill

TECHNIQUES

Submersion dyeing

Steaming

Wirework

Soldering

Sawing or using a disk cutter

Finishing

Roller printing

Drilling

PROCESS

1. Use the permanent-marking pen to draw on the anodized aluminum, creating a pattern of your choice. The charms will move and flip over the bracelet, so draw a different pattern on each side of the metal.

2. Gently lower the aluminum into a dye bath. Leave the metal submerged until the desired color is obtained.

3. Thoroughly rinse the dyed metal. Gently rub it with pumice powder or a kitchen scrub pad to remove any excess ink from the pens. Steam the aluminum for approximately 40 minutes to seal the color.

4. Use the 0.8-mm silver wire and the small mandrel to make 18 oval jump rings, each 1 cm long. Link and solder the rings together, leaving one end open for the bar clasp. This is the bracelet chain.

5. Center a 4-mm jump ring on one side of the 2-mm silver wire. Solder the jump ring to the wire. This is the bar clasp.

Charm Bracelet

6. Using pliers, join the bar clasp to the open end of the bracelet chain. Solder the open link closed, then pickle and rinse the bracelet.

7. File off any sharp edges or excess solder. Polish the bracelet with a silver cloth or in a tumbler.

8. Retrieve the colored aluminum from the steamer, being very careful not to scald yourself. Tip: Lift the lid away from you and wait a second to let the steam clear. Rinse the sealed aluminum and rub it gently with pumice or a kitchen scrub pad to remove any scum that may have built up on the surface. Dry the metal thoroughly.

9. Determine the most interesting or colorful sections of the aluminum sheet. Using a jeweler's saw or a disk cutter, cut 10 round disks from these sections, each 20 x 12 mm. Cut five more disks, each 10 x 9 mm. Note: It may be easier for you to first cut your sheet into smaller, more manageable sections before cutting out the disks. File and finish all cut edges.

10. Place a piece of lace over an aluminum disk. Run this stack through a rolling mill to texture the metal. Repeat this step for the remaining aluminum disks.

11. Select a drill bit that is slightly larger than the thickness of the jump rings. Drill each oval charm 3 mm from one end. De-burr the drill holes by twisting a larger bit over the hole.

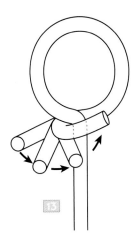

12. Pair each of the small charms with a large charm. Thread each pair onto a heavy 5-mm jump ring. Thread each of the remaining large charms onto a jump ring. Attach the charms to the bracelet, alternating between paired charms and single charms and spacing the charms evenly apart down the length of the chain.

13. Thread a large bead then a smaller bead onto a head pin. Using round-nose pliers, bend the head pin to one side just above the smaller bead and begin to twist a loop. Thread the loop onto the third link of chain from the bar clasp. As shown in the illustration, wrap the excess headpin wire around the base of the loop to secure.

14. Repeat step 13, alternating beads and aluminum charms until you have completed each link in the chain. Leave the last link plain so the clasp will have room to move.

Crosshatch Brooch

This brooch features a simple mechanism that requires no hot connections. Incorporating linear accents throughout the design, I have created my own crosshatch pattern, using a simple print technique and tropical hues.

PROCESS

1. Mix two hues of orange ink for this project: one from yellow and red ink only and a paler version created by adding clear to the yellow and red mix.

2. Spread some paler orange ink into a shallow container, such as a jam jar lid. Dip the edge of the flat applicator into the ink. Carefully and methodically apply the pale orange ink to the anodized aluminum sheet in one direction, picking up ink all the way and covering the entire sheet with hatch marks.

3. Turn the aluminum sheet 90 degrees and repeat step 2 (see photo). Wipe the flat applicator clean.

4. Repeat steps 2 and 3, this time using the pure orange ink. Let the inks dry thoroughly on the metal.

5. Submerge the aluminum sheet in a shallow bath of turquoise water-based

Crosshatch Brooch

dye. Pour a drop of yellow dye over the metal sheet. Move the bath so the dyes mix together. Rinse the metal and steam it to seal the dye.

The yellow dye is absorbed into the aluminum sheet in the area where it was poured. This creates a great blended patch with no edges or marks. As the yellow dye continues to mix with the turquoise dye, the rest of the sheet takes on a diluted version of the yellow, resulting in a beautifully exotic sea green.

6. Use a solvent to completely remove the orange ink from the sealed aluminum surface.

7. Measure and mark a shape on the aluminum with two 6-cm sides, one 2-cm end, and one 3.5-cm end. Use a jeweler's saw to cut out the marked shape. File and finish the metal with smooth beveled edges and rounded corners to nicely frame the pattern.

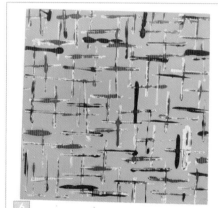

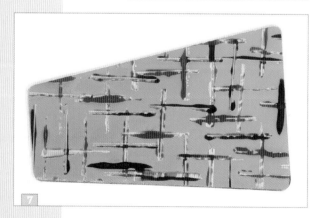

8. Measure and mark a centered point on each narrow end of the aluminum, 1 cm from the edge. Using a 0.8-mm bit, drill holes at the marked points. Remove all burrs from the drilled holes.

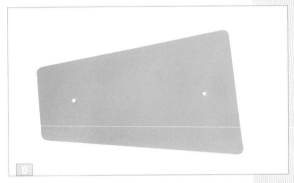

9. Using a cross-peen hammer, strike lines into the design on the aluminum surface, making sure to add this texture in both directions.

10. To make the pin finding, clean one end of the 10-cm wire with a file. File the other end of the wire to a point. Measure 1 cm from the flat end of the wire, and bend a right angle at this point. Measure 4 cm from the first bent angle, check that this spacing lines up with the drilled holes, and bend another right angle in the same direction.

11. Starting with the longer end, feed the bent wire through the drilled holes. Using round-nose pliers, carefully bend a loop at the shorter wire end to make the catch, leaving a small gap for the end of the pin to slot through.

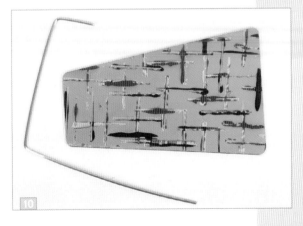

12. Measure approximately 5 mm up from the base of the pin stem. Make another tightly angled bend at this point so the pin runs parallel to the back of the brooch. Secure the pin stem in the catch.

Crosshatch Brooch **71**

Stamped Cuff

Adding texture to printed aluminum lends a completely new dimension to your design. Layering colors and patterns with texture creates interesting depth and offers infinite possibilities of combinations.

YOU WILL NEED

Anodized aluminum sheet, 1 mm, 5 x 15 cm, preferably annealed

Medium mesh, brass, copper, or steel, 2 x 17 cm

Inks, turquoise and pale aqua/green

Water-based dye, brown/bronze

Dyeing kit, page 23

Basic tool kit, page 24

Stamps of your choice

Mandrel or wood dowel, 1½ inches (3.8 cm) in diameter

Rawhide mallet

TECHNIQUES

Stamping

Submersion dyeing

Steaming

Roller printing

Filing

Hammering

Finishing

PROCESS

1. To create the first layer of color, use a paintbrush to cover the stamp with pale aqua/green ink.

2. Firmly press the inked stamp onto the aluminum, and carefully lift it off to prevent smudging. Continue this process to cover the aluminum as desired.

Stamped Cuff

3. Using either a new stamp or the same one (cleaned with acetone), repeat steps 1 and 2 with turquoise ink. Leave to dry.

4. Overdye the stamped aluminum sheet in the brown water-based dye. The stamped area will remain the original color. Rinse and steam the stamped and dyed metal.

5. When the steaming process is complete, remove the ink from the aluminum with thinner and thoroughly dry the metal.

6. Cut the mesh to size and place it on the aluminum sheet.

Tip: If you're concerned about the mesh moving when it's rolled through the mill, cut a piece of paper the same size as the aluminum sheet and use double-sided tape to secure the mesh to the paper.

To check that the rollers are set to the right height for a successful print, make a test run with a piece of scrap aluminum and a small section of mesh. When you're happy with the results, you're ready to proceed.

7. Carefully align the mesh and the edge of the aluminum with the rollers. Steadily turn the arm on the mill until the sheet comes through the other side. Remove the squashed mesh.

8. Using a fine or medium file, bevel the edges of the aluminum to a smooth finish. File each corner into an even curve. Finish all the metal edges and corners with an emery paper until smooth.

9. Place the mandrel in a vise horizontally. Using a rawhide mallet, start to bend each end of the aluminum (photo A). Work steadily to prevent creating folds. Continue hammering until you achieve a smooth curve in a slight oval formation with a gap of approximately 1 to 1½ inches (2.5 to 3.8 cm) as shown in photo B.

10. Use pumice or a kitchen scrub pad and liquid dish soap to clean off any marks left by the mallet on the cuff.

Necklace

Adaptable at every stage, this necklace project gives you the chance to personalize your final outcome. It uses simple materials, minimal tools, and just two colors of dye. The colors shown mix well, lending themselves to layering and relief work that changes quite dramatically as you progress.

YOU WILL NEED

Anodized aluminum sheet, 1 mm, at least 6 x 13 cm

Sterling silver round wire, hard, 1 mm, 80 cm long

2 water-based dyes, turquoise and violet

Dyeing kit, page 23

Basic tool kit, page 24

Soldering kit, page 25

Masking tape

TECHNIQUES

Submersion dyeing

Wirework

Balling wire

Sawing or cutting

Finishing

Drilling

AUTHOR'S NOTE

The dyeing process for this project is additive and uses colors that blend together well. If you wish to use opposing colors, simply alter the process to be subtractive, using jeweler's pickle to bleach out the color from unmasked areas at each stage.

PROCESS

1. Dye the anodized aluminum sheet in the water-based turquoise dye. Rinse and dry the metal thoroughly.

2. Tear masking tape into small pieces with random shapes, each approximately 2 cm. (Tip: Remove your gloves to do this and stick just part of the torn pieces of tape onto a heavy object. Leaving an edge free helps you pick up the tape later.) Tearing the masking tape creates more organic edges. Cutting the tape with scissors or a knife is also fine. This allows you to achieve a more defined edge.

3. Randomly arrange the pieces of tape onto the aluminum sheet. Apply them lightly and not too far apart from each other. Take care to avoid skin contact with the metal and to rub down the edges of the tape. Determine if you would like to have the design on the back of the necklace elements. If so, repeat this step on that side. (In this project, the back does not have the design.)

4. Dye the aluminum sheet in the water-based purple dye very briefly (straight in and straight out) to achieve a bluish purple color. (The exact time depends on the strength of your dye.) Rinse and dry the metal thoroughly. Remove the tape pieces to reveal the turquoise below.

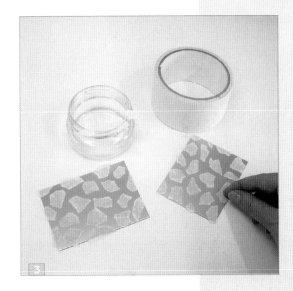

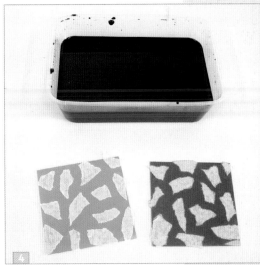

Necklace

5. Repeat step 3, making sure to place pieces of tape in different positions and to cover the turquoise surface in parts.

6. Dye the aluminum sheet in the water-based purple dye for a second time, keeping it submerged longer to achieve a brighter hue. Rinse and dry the metal thoroughly. The residue from the first set of tape you removed will have slightly stopped out some of the second dye bath, adding yet another color.

7. Remove the tape pieces to reveal the finished pattern, consisting of at least three colors—turquoise and two different purple hues. Steam the aluminum to seal the dye for at least 40 minutes.

8. Cut the sterling silver wire into 10 pieces: four 7-cm lengths, five 8-cm lengths, and one 9-cm length. (The 9-cm wire will be the catch for the clasp.) Hold one of the wires vertically with metal tweezers, and apply flux to the bottom end. Using a torch, heat the fluxed end to create a ball that is large enough not to slip through a small loop created with the same wire. Pickle the balled wire and polish it with a silver cloth. Repeat this process to ball one end of all 10 wires. Gently curve each wire section.

9. Use a soft abrasive sponge to remove any scum from the sealed aluminum. Wipe away any tape residue with a little solvent or thinner. Dry the metal.

10. Using a jeweler's saw (or snips if your aluminum is less than 0.7-mm thick) cut five 6 x 2.5-cm oval shapes from the dyed aluminum sheet. Finish the cut edges with a file, 400-grit emery paper, or an abrasive burr on a flexible shaft.

11. Measure and mark two centered points on all aluminum ovals, one on each narrow end and 3 mm from the end. Drill a hole at each marked point. Use a small needle file to give the drilled holes an oval shape. (This allows for shallower curves in both the wire and the aluminum.)

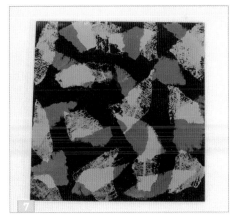

Necklace

12. Using a sandbag and mallet or your hands, gently curve each of the aluminum ovals.

13. Thread the 9-cm piece of wire through the holes in an aluminum oval. Use round-nose pliers to make a very small closed loop at the un-balled wire end.

14. Bend an open loop on the un-balled end of a 7-cm wire. Thread this end over the 9-cm wire near the balled end. Close the loop to secure. Repeat this process, alternating 7-cm wires and 8-cm wires and using the aluminum ovals on every 8-cm piece. Note: The last wire used will be 8 cm long, but it will not have an aluminum oval.

15. Bend the balled end of the first wire back on itself to make a hook. Fit the hook through the loop at the opposite end of the necklace to finish the clasp.

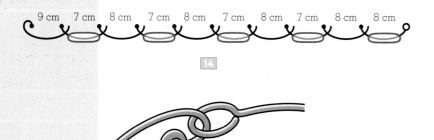

Wire Studs

Wire Studs

Because of aluminum's inability to be easily soldered, designing and making stud earrings is often difficult. This project shows an alternative method of making studs that you can adapt to any style.

YOU WILL NEED

Anodized aluminum sheet, 1 mm, at least 6 cm square

2 sterling silver ear posts, 0.8 mm, each 1 cm long

2 sterling silver wires, 1 mm, each 2 cm long

2 ear nuts

Ink, blue, red, and clear

Water-based dye, olive green

Dyeing kit, page 23

Basic tool kit, page 24

Soldering kit, page 25

Tissue paper

Drill bit, 0.8 mm

TECHNIQUES

Mixing ink

Stamping

Submersion dyeing

Sawing

Filing

Finishing

Drilling

Wirework

Soldering

PROCESS

1. You will need two hues of purple ink for this project. For one, mix red and blue ink to make a deep purple. For the second, mix some of the purple ink with clear ink to achieve a paler hue.

2. Scrunch up a small piece of tissue paper to use as an applicator. Use this to randomly apply the paler purple ink to the aluminum sheet. (To make sure you always have the right amount of ink on your applicator, dab it on a tissue each time you pick up more ink.) Cover the entire aluminum surface.

3. Repeat step 2, this time with the darker purple ink. Leave areas free for the background water-based dye.

4. Using the end of a paintbrush, apply small spots of the darker purple ink that are a couple of centimeters apart. Let the ink dry completely.

5. Submerge the aluminum sheet in the olive green water-based dye. Rinse the metal and steam it to seal the color.

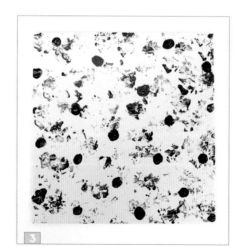
3

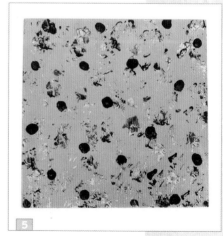
5

6. Measure and mark two 1.5-cm squares on the aluminum sheet. Use a jeweler's saw to cut out both marked squares. File each edge to make a pillow shape. File a gentle beveled edge on both forms and finish with emery paper.

7. Find the center point on each metal shape by drawing a diagonal line from corner to corner. Indent each center point with a center punch and drill it with an 0.8-mm bit.

8. Bend a gentle curve in each 2-cm piece of silver wire. Solder an ear post at a 90-degree angle to the center of each bent wire. Pickle and clean the wires.

9. Thread an ear post through each hole in the aluminum forms, and secure with an ear nut.

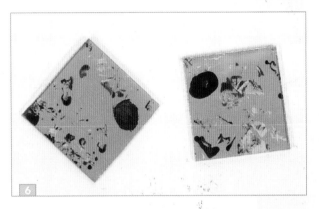

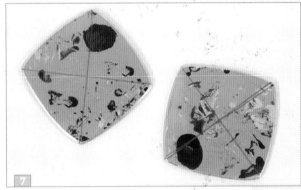

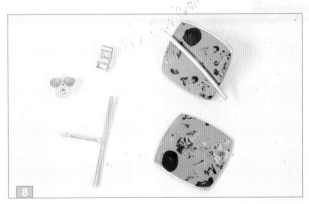

Leaf Earrings

These delicate drop earrings are made from thinner aluminum sheet to allude to their natural source of inspiration. Printing only one side of the sheet provides glimpses of contrast between the interior and exterior. The different angled lines provide interest whichever way the earrings hang.

YOU WILL NEED

Anodized aluminum sheet, 0.5 mm, at least 8 x 8 cm

2 sterling silver wires, 0.8 mm, 4 cm long

Ink, yellow

Water-based dye, brown

Dyeing kit, page 23

Basic tool kit, page 24

Soldering kit, page 25

2 ink rollers

Thin acetate

Drill bit, 0.8 mm

Forming block or small mandrel

TECHNIQUES

Rolling ink
Monoprinting
Submersion dyeing
Sawing
Finishing
Drilling
Bending
Wirework

PROCESS

1. Roll a small amount of yellow ink onto the acetate. Make a thin, even layer that is approximately 10 cm square.

2. Use a pencil to draw quick vertical lines in the left section of the inked area. On the right section, draw horizontal lines. Make sure the lines meet in the middle of the inked area. By drawing the lines, you're removing the ink from the acetate plate.

3. Taking care not to disturb the ink, pick up the acetate, line up the pattern over the aluminum, and place the acetate onto the metal. With a clean roller, roll over the acetate with smooth, even movements.

Leaf Earrings

4. Carefully peel away the acetate to reveal the printed image. Let the ink dry on the metal. Thoroughly clean the acetate with solvent until no color remains.

5. Submerge the aluminum sheet into the brown water-based dye until the desired color is achieved. Remove the ink with thinner. Rinse the metal and steam it to seal the color.

6. Draw two double leaf shapes on the dyed aluminum. Use a jeweler's saw to cut out the marked shapes.

7. Finish the edges of the aluminum double-leaf shapes with emery paper.

8. Mark the center of the middle section of each double-leaf shape. Using a 0.8-mm bit, drill a hole at each marked point.

9. Curve each double leaf up to the middle section. If you have a forming block, use it with the side of a covered punch (see photos A and B). Tap the metal lightly and adjust the depth to prevent marking any edges. If you don't have a forming block, just lightly hammer the aluminum around a small mandrel.

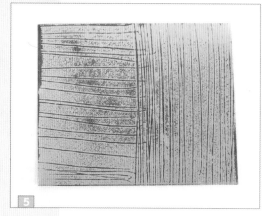

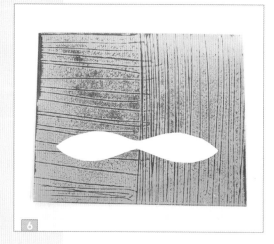

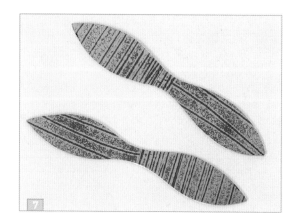

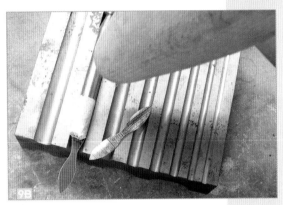

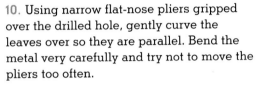

10. Using narrow flat-nose pliers gripped over the drilled hole, gently curve the leaves over so they are parallel. Bend the metal very carefully and try not to move the pliers too often.

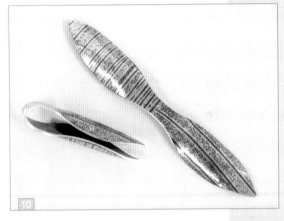

11. For the ear wires, you'll make a ball at one end of each 4-cm piece of silver wire. Flux the wire end, hold it vertically, and heat the bottom with a torch until a ball forms. Pickle and clean the balled wires. File the end of each wire to take away any sharp edges.

12. Thread each wire through the drilled hole in an aluminum leaf. (The balled wire end acts as a barrier to prevent the leaf from dropping off the wire.) Measure 1 cm up the wire from the top of the leaf. Using large round-nose pliers, bend the wire around and down, taking care to evenly curve both ear wires. Make a gentle bend away from the leaf to finish the shape of the ear wires.

Ring

Practice three important skills—image transfer, simple cold forming, and tube riveting—in a single project. A photocopy and acetone are used to transfer the design, but the acetate or PNP paper methods would work equally well.

YOU WILL NEED

Anodized aluminum sheet, 1 mm, 3 cm square*

Sterling silver round wire, 2 mm, approximately 6 cm

Sterling silver tube for rivet, 2.2 mm outside diameter, 1.4 mm inside diameter

Water-based dye, turquoise

Water-based dye, yellow

Photocopied design template (below, right) or image of your choice to fit your sheet

Dyeing kit, page 23

Basic tool kit, page 24

Soldering kit, page 25

Masking tape

Kitchen towel or cotton cloth

Acetone or acetone-free nail-polish remover

Ring mandrel

Rounded wooden mallet or wooden doming punches

Sandbag or wooden doming block

*A 3-cm square is the size of the final piece, but you may wish to use a larger sheet of aluminum. This would allow you to transfer multiple images so you can choose and cut out the best one. A larger sheet also makes it easier to place and transfer your image. You can also adapt the project to feature a larger image if you have a larger sheet.

TECHNIQUES

Submersion dyeing

Photocopy transfer

Forming

Soldering

Sawing

Filing

Drilling

Riveting

Burnishing, optional

VARIATIONS

Chose an image or pattern to suit your taste and the style of your work, or adapt a digital photograph by adjusting the contrast to high to gain only lights and darks. Whatever image you use, remember to reverse it, especially if it contains writing.

To achieve contrasting colors, for example turquoise and red, remove the turquoise dye after step 5 by dipping the aluminum in jeweler's pickle. Rinse the metal, dye it red, and complete the remaining steps.

PROCESS

1. Dye the anodized aluminum sheet in the water-based turquoise dye. Rinse and dry the metal thoroughly.

template

Ring

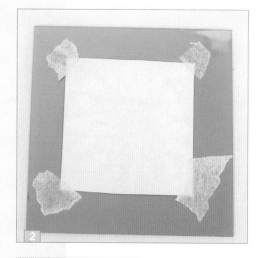

2. Place the photocopied design template on the surface of the dyed metal with the toner facing down. Use small pieces of masking tape to secure each corner of the paper template. Remember: The area where the tape is placed will not be dyed!

3. Lightly dampen a kitchen towel or a cotton cloth with acetone or acetone-free nail-polish remover. (It's important that the cloth is not drenched in thinner. Too much thinner will create a big smudge instead of a well-defined image and can also block the pores of the anodic film, preventing the addition of a second dye.) Let the thinner soak into the cloth for a couple of seconds, then very carefully and evenly dab it onto the back of the paper template. As shown in the photo, the toner image will start to appear through the paper gradually, not immediately. (If you have used too much thinner, wait a few seconds for the paper to dry before moving to the next step.)

4. Gently, thoroughly, and evenly rub the paper surface, taking care not to disturb its placement. Moving the paper will result in a blurred double image.

5. Once the paper is thoroughly dry, carefully lift one corner to check that the image has transferred completely. If the transfer is successful, gently remove the paper. If not, you may have not rubbed the template enough. Continue rubbing the paper, then recheck the image.

6. Over-dye the anodized sheet with the second dye color. (For this project, I chose yellow. As shown in the photo, when yellow is combined with the turquoise, a bright lime green is created. Note: Yellow dyes tend to be fairly powerful. If mixed too strong, they will overwhelm other colors. To make sure your mix is not too strong, do a test run on a practice sheet. Over-dye a turquoise sheet for short periods until you achieve the desired color.) Rinse the metal thoroughly and steam it for 40 minutes to seal the color.

7. Using a ring mandrel and rawhide mallet, bend the silver wire into a band. Solder the ends of the wire to form the ring.

8. Use a jeweler's saw to cut the silver tube slightly longer than the thickness of the aluminum sheet. File each end of the tube. Solder the tube perpendicular to the ring at its solder seam. Pickle, file, and polish the ring. (Heating anneals the tube, which is necessary for making a rivet.)

Ring

9. Using acetone, remove the toner from the sealed aluminum to reveal the turquoise image below (see photos A and B). Use a mildly abrasive sponge to remove the acetone and scum from the aluminum surface.

10. Use a jeweler's saw to cut out the aluminum shape. File the tips of the design to slightly soften

them. Finish all edges with emery paper or an abrasive burr on a flexible shaft.

11. Place the aluminum face up on a leather sandbag. Using a rounded mallet or wooden punch, hammer the metal to create a shallow dish. Just a few blows will result in a suitable curve. Excessive hammering could create cracks in the surface of the metal, but these can be used decoratively.

12. Mark the center point of the aluminum. Center punch the metal at this point. (If desired, cover the metal with masking tape for protection, and rub down the center point so you can see the indent). Select a drill bit that is the same diameter as your tubing. Drill the aluminum and remove the burr.

13. Place the silver ring onto a mandrel and secure the mandrel in a vise. Slide the aluminum onto the tube rivet.

14. With a scribe or center punch, flare out the end of the tube rivet to create a lip. Using a small hammer, lightly tap down on this lip until the aluminum is secure. If necessary, use a burnisher to tidy up any marks.

Rose Necklace

This project takes you all the way from photographing an image to printing the picture onto aluminum. Simple image-editing software programs are used to alter the photo to a fully contrasted image. A silver chain and beads complement and bring color to the rest of the necklace.

YOU WILL NEED

Anodized aluminum sheet, 1 mm, approximately 8 cm square

Sterling silver wire, 1.5 mm, 1 m long

Sterling silver wire, 2 mm, 50 cm

Sterling silver wire, 0.5 mm, 50 cm long

Rose

3 beads, each 1 cm

2 freshwater pearls, each 5 mm

3 freshwater pearls, each 8 mm

Water-based dye, brown/yellow, diluted

Dyeing kit, page 23

Basic tool kit, page 24

Soldering kit, page 25

Digital camera

Computer with image-editing software

Computer printer

Photocopier

Photocopy acetate

Iron

Hammer, medium size

Tumbler or silver polishing cloth or dip

Drill bit, 0.8 mm

TECHNIQUES

Digital photography

Basic computer photo-editing

Submersion dyeing

Acetate image transfer

Sawing

Finishing

Wirework

Soldering

Hammering

PROCESS
Creating the Pendant

1. Place the rose on a white background, and take a digital photograph.

2. Upload the photograph into an image-editing program so you can make adjustments. First, change the color mode to grayscale, which will make your image black and white. Using the brightness and contrast adjustment tool, remove all the halftones (greys) from the image to leave a purely black and white image.

3. Invert the image. (This means turning the blacks to white and the whites to black.) The black will stop out an additional color, leaving the exposed metal (white) sections to be replaced by a second dye. Print the edited image onto paper, and then photocopy it onto acetate.

Rose Necklace

4. Submerge the anodized aluminum sheet in a very pale brown dye to produce a cream-colored background. Rinse the metal, then dry it thoroughly.

5. Place the acetate image, ink side down, on the dyed aluminum sheet. Secure the acetate to the metal with tape. Set the iron on a medium-low heat. Gently and methodically press the acetate with the iron until the ink has transferred to the metal. Let the metal cool. Remove the acetate to reveal the transferred image.

6. Submerge the aluminum sheet in the pink dye and leave it until you achieve the color depth desired. (My dye is quite strong, so this piece only took about 20 seconds to dye.) Rinse the metal and steam it to seal the color. Use a solvent to remove the ink from the sealed metal.

7. Use a jeweler's saw to cut around the outside of the image. File and finish the metal edges with emery paper.

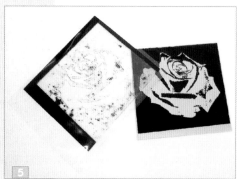

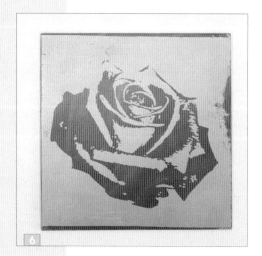

Constructing the Chain

1. Make 16 jump rings from the 1.5-mm round wire, each 1 cm in diameter. Make seven jump rings from the 2-mm round wire, each 1.5 cm in diameter (see photo). This design features a mixture of round and oval links. Solder the jump rings closed, then pickle them. Flatten each ring with a heavy hammer.

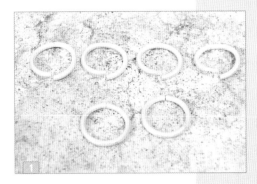

2. Make the link from 1.5-mm wire, using round-nose pliers to form a figure eight and leaving one end slightly open. Solder the closed end of the figure-eight shape. Pickle the wire, then hammer it flat.

3. Lay out all of the components in the necklace design and arrange them to your liking. I've used glass beads and freshwater pearls to bring some of the color of the rose into the chain and to complement the color of the dye.

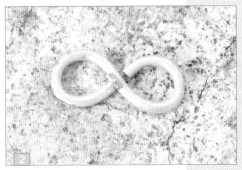

4. Saw apart half of the silver links so the chain sections can be connected. Once connected, solder these links closed. Pickle the chain sections, then place them in a tumbler to bring them to a high shine. If you don't have access to a tumbler, you can achieve a good shine with a silver polishing dip or cloth.

5. Determine and mark two points where you would like to attach the rose to the chain. Drill a hole at each point with a 0.8-mm bit. Thread a 4-mm jump ring through each hole in the rose.

6. Using figure 13 on page 67 as a guide, thread the beads onto 0.5-mm silver wires, attach each end to a chain link or jump ring, and make a wrapped loop to secure this connection. Repeat this process until all necklace sections are joined.

Disk Pendant

Graphic imagery and simple forms are a reflection of my own jewelry style. This pendant shows how pure image, straightforward form, and clean connection can be an effective way to design and wear jewelry for maximum impact.

PROCESS

1. Photocopy the design template onto press-and-peel paper, or scan the template into your computer and print it on the PNP. Cut out the copied or printed design, leaving a generous margin.

2. Secure the design, matt side down, on the anodized aluminum sheet. Press the design with an iron until the image turns very dark. This may take up to two minutes. Allow the metal to cool completely.

3. Peel off the PNP paper, leaving the transferred design.

4. Dye the anodized aluminum sheet in the water-based red dye. Rinse the metal and then steam it to seal the color.

5. Soak a kitchen towel with solvent, and use this to remove the PNP resist from the fully sealed aluminum.

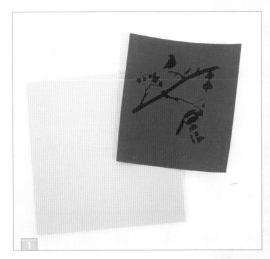

template

Disk Pendant **99**

Disk Pendant

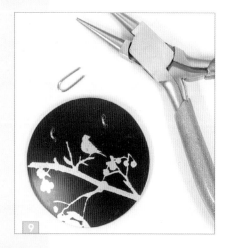

6. Measure and mark a circle around the image that is approximately 4 to 5 cm in diameter. Use a jeweler's saw to cut out the aluminum disk. File and finish the cut edges.

7. At the top of the disk, measure and mark the locations for two pairs of holes (see photo for approximate placement). Center punch the marked points, drill the metal with a 1-mm bit, and remove all burrs.

8. Place the aluminum disk on a sandbag with the printed image facing down. With a round wooden mallet, gently hammer the metal to create a shallow dome.

9. Using round-nose pliers, bend both lengths of the stainless steel or hardened silver wire in the center until the ends are 4 mm apart.

10. From the front side of the aluminum disk, thread each piece of wire through a set of drilled holes. From the back of the disk, bend each wire back on itself. This creates the loops for the neck wire. Thread the pendant on the neck wire to complete the project.

Forged Bangle

Forged Bangle

There is a very free and fluid nature to the design of this bangle. Forging the wire creates depth and texture, and pale ink highlights these unique dimensional qualities.

YOU WILL NEED

Aluminum round wire, 4-mm gauge, 70 cm

Inks, turquoise and clear

Water-based dye, teal

Dyeing kit, page 23

Basic tool kit, page 24

Soldering kit, page 25

Assorted hammers, such as ball peen and cross peen

Print roller or sponge

Acetate

TECHNIQUES

Filing

Forging

Annealing

Wirework

Anodizing or access to commercial anodizing facility

Inking

Submersion dyeing

AUTHOR'S NOTES

This project requires anodizing the aluminum after forming it.

The aluminum wire will lengthen in the forging process.

PROCESS

1. File one end of the aluminum wire to a point. (This helps prevent the wire end from folding in when forged.)

2. Using a hammer and an anvil or steel bench block, start drawing down the wire in a gradual taper until the tip is half the size of the original wire gauge or thinner.

3. Form different textures along the entire wire length by adjusting your tool or technique roughly every 8 cm (see photos A and B). For example, change from round to square shaping, small to large hammers, ball-peen to cross-peen heads, soft to sharp marks, and thin to thick sections. Although there are no right or wrong shapes and textures, if you haven't forged before, you might find it helpful to practice these techniques in advance.

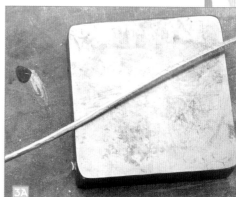

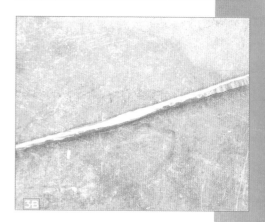

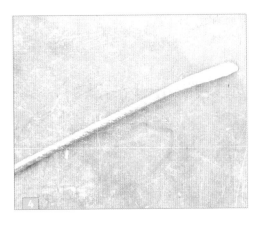

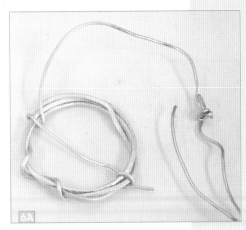

4. As you near the end of the wire, use the hammer to begin spreading it out. Blend this spread into the last texture made. The hammering is now complete.

5. Bend the wire into a coil made of three complete circles, intertwining the strands as desired. Anneal the metal as needed, keeping a careful eye not to overheat the ends.

6. When the metal has cooled, finalize the shape, size, and twists in the bangle design. Because the metal should be really soft, you can use your hands to do most of the work. A pair of pliers may be helpful for tightening the last twist around the bands. There is no right or wrong pattern. Just keep adjusting the wire until you are happy (photos A and B).

Forged Bangle

7. Anodize the forged aluminum bangle (see page 12 for details on this process or take your work to a commercial anodizing facility).

8. Spread the pale turquoise ink on an acetate sheet (photo A). Using a print roller or sponge, apply the ink to the higher points of texture on the bangle (photo B). Color both the inside and outside of the bangle, replenishing the ink often until all areas are covered. Let the ink dry.

9. Submerge the bangle in a teal water-based dye until the desired color is achieved. Rinse the bangle and steam it to completely seal the color. Use a solvent to completely remove the dry ink.

Forged Hair or Sweater Pin

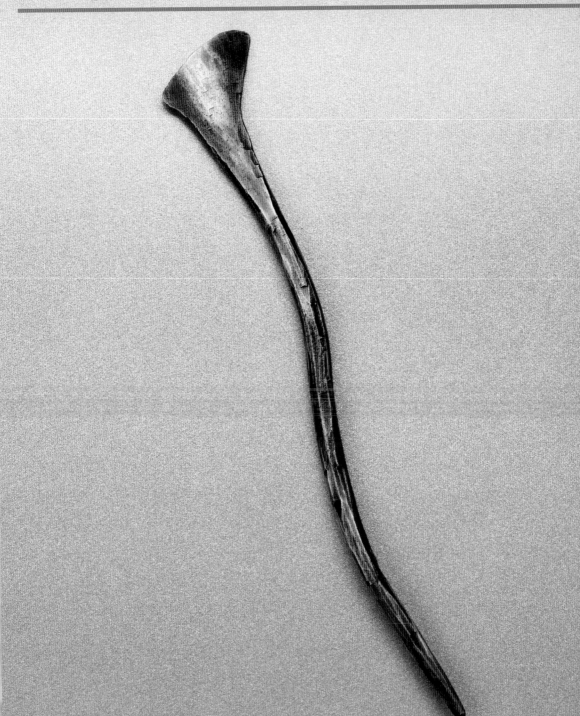

Forged Hair or Sweater Pin

This multipurpose project is a simple example of cold-forging aluminum that lets you experience the wonderful malleability of this metal. It combines the heavy hand of the hammer to create a jagged surface for gripping hair with the gentle touch of the paintbrush to blend the dyes.

YOU WILL NEED

Soft round aluminum rod, 10 mm, 14 cm long*

2 water-based dyes, red and yellow (or colors of your choice that blend well)

Dyeing kit, page 23

Basic tool kit, page 24

Soldering kit, page 25

Cross-peen hammer

Plastic paint palette

Jar of water and rinse bath or nearby water faucet

* I use armature wire and purchase it from a sculpting supplier.

TECHNIQUES

Filing

Forging

Annealing

Anodizing or access to a commercial anodizing facility

Painting with dye

AUTHOR'S NOTES

This project requires anodizing the aluminum after forming it.

The secret of success with this project is annealing! Periodically heating the aluminum during the forging process makes the metal easier to form.

At first, brush blending the dyes may seem difficult, but you'll soon get used to each step and fly through the process with ease.

PROCESS

1. File one end of the aluminum rod to a gentle point. This prevents the soft aluminum from folding into the end when you hammer it.

2. Begin to hammer the pointed end of the rod with force, using the flat, round end of the hammer and continually turning the rod to keep the section rounded.

3. As the pointed end of the rod starts to narrow, begin hammering up the rod with even blows until you finish three quarters of its length.

4. Hammer the non-pointed end of the rod without turning it to create a V-shaped flat area that points down to the previous hammering (photos A and B).

5. Anneal the aluminum rod with a gas torch and let it cool naturally.

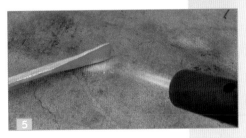

Forged Hair or Sweater Pin

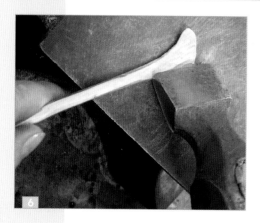

6. Using the wide end of the hammer, start to splay the flattened end of the rod like a fan, keeping one corner of the hammer stationary but changing the horizontal angle of the hammer blows (see figure 1). File the end to create a diagonal. If needed, anneal the rod during this process to prevent any cracks from occurring.

7. Starting from the pointed end of the rod, keep hammering and twisting the metal, but this time use the hammer at an angle to create ridges on the surface. The rod will naturally start to twist and bend into an S shape. Continue this process until you achieve a pin shape that pleases you.

8. Anodize the forged aluminum pin (see page 12 for details on this process or take your work to a commercial anodizing facility).

9. Pour the yellow and the red water-based dyes onto a plastic paint palette. Make sure that your water supply and paper towels are close at hand.

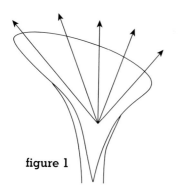

figure 1

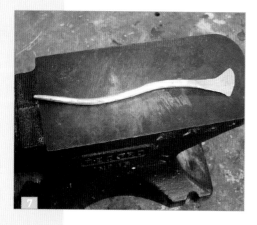

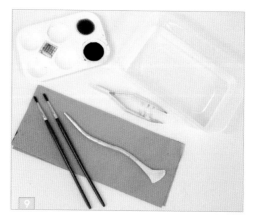

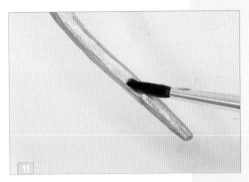

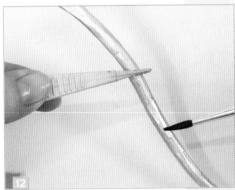

10. Hold the anodized aluminum pin in plastic tweezers with the splayed end down. With a small artist's brush, carefully paint a small section of the splayed area yellow, taking care not to apply too much dye or it will run. (Tip: If you do make a mistake, dip the aluminum into pickle to remove the dye, but make sure to thoroughly rinse and dry the cleaned metal.) Rinse and dry the pin.

11. Turn the pin so the pointed end faces down. Water down a small amount of yellow dye, and use the medium brush to paint up from the tip to the upper middle section of the pin.

12. Gradually build up the color towards the tip until this area is absolutely yellow. Rinse and dry the pin thoroughly.

13. Keeping the tip of the pin facing down, carefully paint diluted red dye around but not on the yellow area at the top of the pin and work down, increasing the strength of the dye over the entire surface. Make sure to move the tweezers regularly to prevent marks.

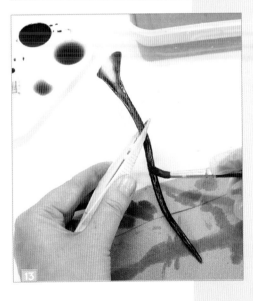

14. Once you achieve the desired color intensity, rinse, dry, and steam the pin.

Knot Earrings & Brooch

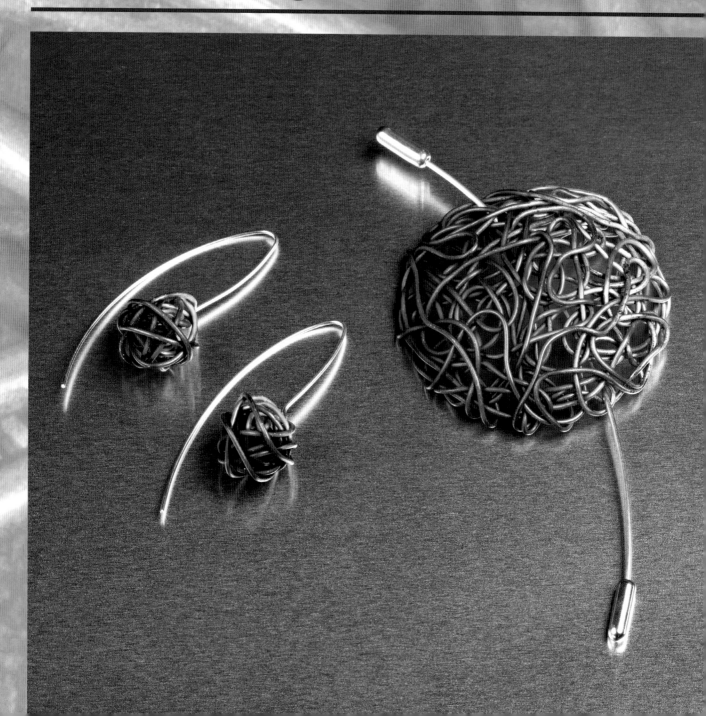

While creating this project, you'll develop skills you can adapt to create any number of knotted items. It may take a bit of practice to achieve the "random-yet-even" quality needed for a successful look, but once you catch on, you'll find it really very easy!

YOU WILL NEED

Aluminum round wire, 0.8 mm, at least 4 m

2 sterling silver round wires, hard, 0.8 mm, each 7 cm long

Sterling silver round wire, hard, 1 mm, 9 cm long

2 stickpin clutches or pin protectors

Water-based dye, blue

Dyeing kit, page 23

Basic tool kit, page 24

Round wooden mallet

Sandbag

TECHNIQUES

Wirework

Anodizing or access to a commercial anodizing facility

Painting with dye

Filing

AUTHOR'S NOTES

This project requires anodizing the aluminum after forming it.

The knotted wire brooch is easy to convert into a pendant and back into a brooch, giving you a choice of how to wear it.

PROCESS

For the Earrings

1. Using round-nose pliers, make a small loop in the aluminum wire.

2. With round-nose pliers, continue to curve the aluminum wire, forming a new full loop around the outside of the original loop.

3. Change the direction of the wire and curve it around the second loop, taking care not to create bends and corners.

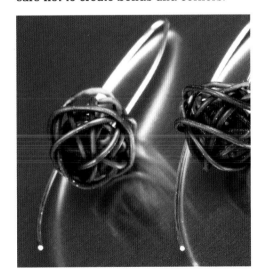

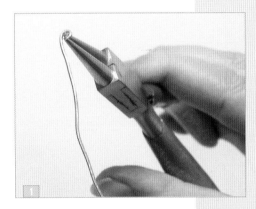

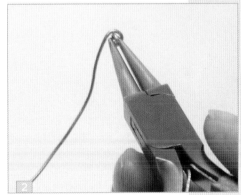

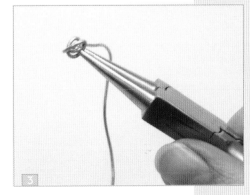

Knot Earrings & Brooch

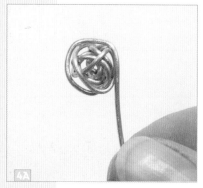

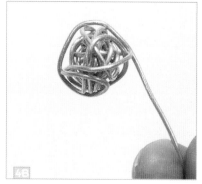

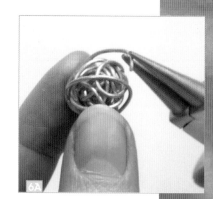

4. From this point, you may find it easier to bend the wire with your fingers rather than pliers. Continue curving the wire and changing its direction (photo A). You will start to see a ball forming (photo B). It may take a few practice runs for you to be happy with the shape of the wire.

5. When the wire knot reaches the desired size, use wire cutters to cut the wire approximately 1 cm away from the knot.

6. Create a small half loop at the end of the wire (photo A). Carefully tuck in this wire end toward the center of the knot so that it is not visible (photo B).

7. Repeat steps 1 through 6 to make a second knot of aluminum wire.

8. Before anodizing, the aluminum wire knots must be securely jigged. I used a thicker wire to make the jig and threaded it between the gaps of the knotted wire. If desired, make and anodize more knots than needed. It's always a good idea to have extras in case you make a mistake.

9. Anodize the aluminum wire knots (see page 12 for details on this process, or take your work to a commercial anodizing facility).

10. Remove the knots from the anodizing jig and place them individually on smaller jigs. These will act as handles for you to hold onto when dying the metal.

11. Use a small paintbrush to heavily apply diluted blue dye to the anodized knot, working from the bottom up (photo A). Keep applying the dye, gradually picking up more dye and applying it to the bottom section. Take care to blend each layer of color into the next. At each of these stages, you'll need to almost "flood" the knot with dye so it can reach the inside of the knot (photo B).

12. For the final stage in the blending process, apply pure undiluted dye to the bottom of the knot. Rinse the dyed aluminum knot and steam it to seal the color.

13. Repeat steps 11 and 12 to dye and seal the second aluminum knot.

14. To make the ear wires, use round-nose pliers to bend each 7-cm wire into a J shape. Bend a small loop at the end of the short side of each J-shaped wire. Smooth the other wire ends with a file or emery paper. Thread the top wire of the knot (the palest part) onto the loop in the ear wire. Close the loop to secure the earring. Repeat this process to connect the second knot to the second ear wire.

Knot Earrings & Brooch

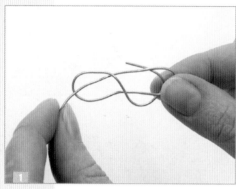

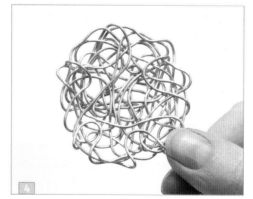

For the Brooch

1. Randomly twist and interweave the wire, bending and curving it to form a disk shape.

2. Continue gradually and evenly building up the size of the wire disk to the desired diameter, keeping the wires fairly loose. Try not to overthink your twisting or the random effect will be lost.

3. Start to build up the edges of the wire disk. Don't cross the center too often. (You are trying to achieve a standard thickness for the whole disk, and crossing the center too often will make it too thick.) Double back the wire and interweave it to help prevent uneven-looking patches. These actions also help to give strength to the structure so it won't easily unravel.

4. When you're happy with the form, cut off the excess wire and securely tuck in the end so it's not visible.

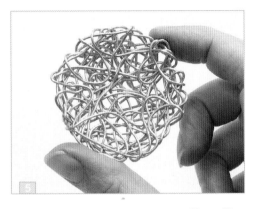

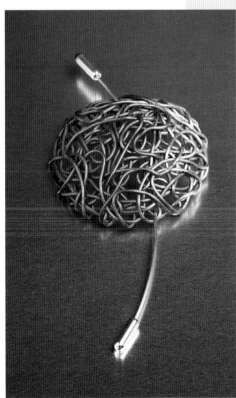

5. Place the wire form on a sandbag. Use a round wooden mallet to hammer the form into a slightly domed shape.

6. Before anodizing, the aluminum form must be securely jigged. I used a thicker wire to make the jig and threaded it between the gaps of the wire form.

7. Anodize the aluminum wire (see page 12 for details on this process, or take your work to a commercial anodizing facility).

8. Remove the wire form from the anodizing jig and place it on a smaller jig. This will act as a handle for you to hold onto when dyeing the metal.

9. Dye and seal the wire form, following the same method used for the earrings (page 113, steps 11 and 12).

10. To make the pin, file each end of the 9-cm silver wire to a point. Bend a gentle curve in the wire with your fingers. Thread the pin through the wire form and slide a stickpin clutch onto each end.

Feather Brooch

Integrating anodized aluminum with conventional jewelry
materials brings a new and colorful element to your work.

Anodized aluminum
sheet, 1 mm, approxi-
mately 60 x 60 mm

Sterling silver sheet,
1 mm, approximately
60 x 60 mm

Sterling silver round
wire, hard, 70 mm

Sterling silver wire,
fully annealed, 30 mm

Silkscreen of a feather
design or other motif

Lime green water-
based dye (7 parts yel-
low, 2 parts turquoise,
1 part purple)

Black anodizing
silkscreen ink

Dyeing kit, page 23

Basic tool kit, page 24

Soldering kit, page 25

Acetone or silkscreen
cleaner in a spray bottle

Packing tape

Newsprint and
paper towels

Heavy natural paper

TECHNIQUES

Submersion dyeing

Silkscreening

Roller printing

Sawing

Drilling

Soldering

Finishing

Riveting

BEFORE YOU BEGIN

Prepare your silkscreen, your silkscreening area, and lay out your equipment
within easy reach. If needed, refer to
pages 58–60 for details of this process.

PROCESS
Dyeing & Printing the Aluminum

1. Dye the anodized aluminum sheet pale
lime green. Rinse and thoroughly dry the
dyed metal.

2. Mask off a frame around the area to
be silkscreened with packing tape, covering the entire area over which the ink
will be drawn (see photo). Place a piece
of newsprint under the screen.

Feather Brooch

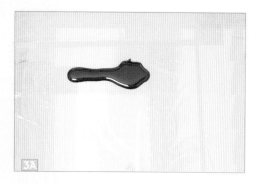

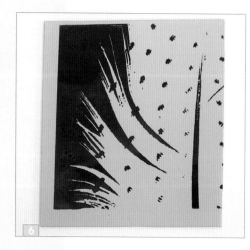

3. Pour a channel of black ink across the top part of the screen (photo A). Secure the screen frame with an extra pair of hands or a clamp. Use a squeegee to steadily pull the ink across the image, "flooding" the screen (photo B).

4. Tilt the screen up, remove the newsprint, and replace it with the dyed aluminum sheet. As you lower the screen, line up the edges of the image within the edges of the metal.

5. Re-secure the screen, then use the squeegee to steadily pull the ink over the image twice.

6. Lift the screen and remove the metal very carefully so the image does not smudge. Place the printed metal somewhere clean and dust free. Let the ink dry for approximately half an hour. Tip: To speed up the drying time, you can use a hair dryer on a low setting.

7. Quickly clean the screen before the ink dries into the mesh. Spray acetone or screen cleaner onto the mesh, and wipe it with paper towels and layers of newsprint. Keep replacing the paper towels and newsprint until the color is no longer present. This can take quite a while!

8. Steam the dry aluminum to seal the dyeing and printing.

9. Soak a paper towel with some solvent. Rub it over the surface of the aluminum to remove the black ink. Keep replacing the paper towel until the ink is completely gone.

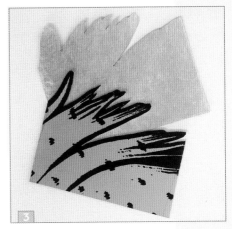

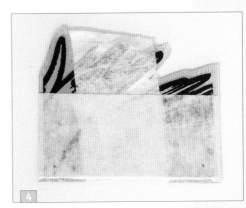

Constructing the Brooch Layers

1. Use a sharp pencil to draw on the printed aluminum, following the outside edge of the feather pattern. Use a jeweler's saw to cut out the pattern along the drawn line. Finish the cut edges with a fine file and emery paper.

2. Place the natural paper on the annealed silver sheet, and roll this stack through the rolling mill to texture the metal.

3. Lay the aluminum feather on top of the silver sheet. Using a pencil or a scribe, draw around the feather, leaving a 2-mm margin. Use a jeweler's saw to cut out the pattern along the drawn line. Finish the cut edges with a fine file and emery paper.

4. Position the two metal layers together, and secure them with tape or a small clamp.

5. Center punch four points on the top metal surface. Select a drill bit that corresponds to the diameter of your rivet wire to achieve a tight fit. (I used a 1-mm bit.) Drill through both metal layers. Remove all burrs.

Feather Brooch

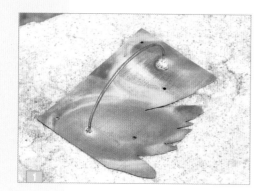

Creating the Pin Findings & Assembling the Brooch

1. To make a simple pin finding, bend the 7-cm wire into a C shape. Solder this wire to the back of the silver sheet approximately 8 mm from its side edges (see photo). Pickle and clean the silver.

2. Near the right soldered point, cut out approximately 10 mm of wire (figure 1). File the end of the short wire length flat. Bend the wire sideways, across the brooch so the end is almost touching the base. This is the catch.

figure 1

3. Near the left soldered point, use round-nose pliers to bend the long wire length into a full loop, and continue bending until the tip of the wire points toward the catch (figure 2). Straighten the stem, line it up with the catch, snip off any excess wire, and gently file its end to a point (see photo).

4. Cut four pieces of annealed, 1-mm wire that are slightly longer than the depth of two brooch layers. Using a small hammer, create a rivet head at one end of each wire.

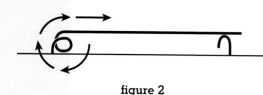

figure 2

5. One at a time, thread a rivet wire through a drilled hole, and hammer the non-flared end to secure the layers. Finish all rivet heads on both sides of the brooch until they are secure and tidy.

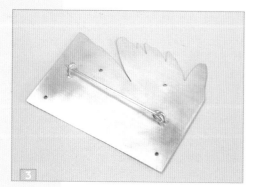

Gallery

Left top: **Lisa Vershbow**
Red Tornado Basket, 2008
27 x 27 x 19 cm
Industrial radiator screen, aluminum, screws, cabinet door pulls; anodized, cut, folded, cold connected
Photo by Studio Munch

Left bottom: **Gayle Friedman**
Knitting Needle Cuff, 2008
3.2 x 7 x 7 cm
Mom's anodized aluminum knitting needles, sterling silver; hand fabricated
Photo by artist

Right top: **Meghan O'Rourke**
Cluster Brooches, 2007
Each, 7 x 6 x 2 cm
Aluminum, titanium, niobium, sterling silver, cubic zirconia, stainless steel; anodized, hand dyed, sandblasted, riveted, bezel set
Photo by Grant Hancock

Right center: **Lisa Vershbow**
Corsages (two brooches), 2008
Largest: 12 x 4 cm
Silver, aluminum; anodized, fabricated, folded, screwed
Photo by Studio Munch

Top left: **John Moore**
Sphere Pendant - Vane Collection, 2008
6 cm in diameter
Aluminum, silver, silicone, neoprene;
anodized, screenprinted, stamped,
fabricated
Photo by Full Focus

Top center: **Mei-Fang Chiang**
Flowered, 2009
18 x 18 x 18 cm
Aluminum, brass; dyed, sawed
Photo by artist

Top right: **Deepti Kumar**
Indigo, Bangle, 2007
8.7 x 1.7 cm
Aluminum; photoetched,
hand painted
Photo by Sussie Ahlburg

Center: **Tracy Rohr**
*6 Disc Bracelets with Multiple
Accents*, 2004
Each, 17.2 x 2.5 cm
Aluminum; riveted
Photo by Taylor Dabney

Left: **Erica Matsuda**
Untitled, 2009
28 x 23 x 2 cm
Aluminum, monofilament; dapped,
anodized, riveted, pierced
Photo by Helen Shirk

Top left: **Lorraine Gibby**
Lace Collar, 2008
4.5 x 38 cm
Aluminum, white precious metal, pearls,
quartz; printed, dyed, textured, linked
Photo by Steve Speller

Top right: **Damia Smith**
Grana Carotene, 2008
22.9 x 33 x 25.4 cm
Aluminum, copper, sterling silver,
mulberry paper; dip dyed, formed
Photo by Annie Pennington

Center: **Niamh Mulligan**
Aloe Vera, 2008
5 x 5 x 1.5 cm
Sterling silver, aluminum; anodized,
hand fabricated
Photo by Norton Associates

Bottom: **Jill Fagin**
Wire Wrap Bangle Bracelets, 2007
Each, 8 x 8 x 3 cm
Aluminum; extruded into wire, hand
wrapped, anodized
Photo by artist

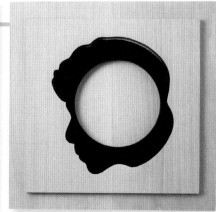

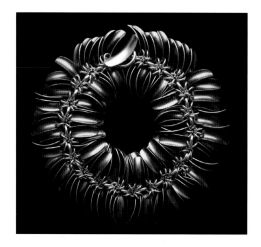

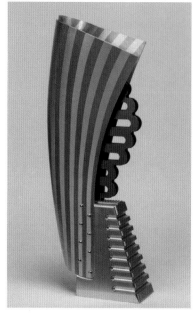

Top: **Kevin Hughes**
Wilma Rudolf, 2007
10 x 8.6 x 0.6 cm
Aluminum; anodized,
hand pierced
Photo by Marc Creedon

Center left:
John Moore
*Charm Bracelet -
Elytra Collection*,
2007
18 cm; each unit, 1.5
x 0.6 cm aluminum;
anodized, dip dyed,
stamped
Photo by artist

Center: **I-Ting Wang**
Blooming IV, 2004
22 x 12 x 4.5 cm
Aluminum, sterling
silver, cubic zirconi-
um; dip dyed, photo
etched, hand painted,
riveted
Photo by artist

Center right: **Moegi Ito**
Wind at the North Pole,
2001
74 x 30 x 16 cm
Aluminum plate, alu-
minum block; beating
metal technique, heat-
ed, annealed, squeezed
by hammer, chemical
polished, anodized dye
patterns
Photo by artist

Right: **Kathleen Keogh**
Key Chain, 2005
10.2 x 1.9 x 53.3 cm
Aluminum, nu gold,
brass; anodized
Photo by artist

Glossary

A

Abrasive – a hard compound, usually mineral, used to shape or finish a piece of work through rubbing

Acetate – thin clear plastic sheet printed and used to create a relief pattern on the surface before or between dyeing

Anneal – heating to soften the material and relieve internal stresses; prevents excess work hardening and stress fractures

Anode – a positive electrode through which current flows into an electrical device

Anodic film or surface – a microscopic pore-like surface or oxide formed by passing a direct current through an electrolytic solution

Anodizing – an electrochemical reaction forming a thin oxide coating on metal

B

Bain-marie – a heated shallow bath of water that makes use of double containers to gently heat its contents

Blanking – a shearing process useful for cutting multiple shapes; a shape is cut into a sheet of steel to form a die (the pattern to be cut); aluminum sheet is placed between the die; pressure is applied by a fly press or sharp hammer blow; the pattern is cut with a scissor-like action

Burr – small drill bit with an abrasive or cutting head, or an overhang of metal caused by drilling or cutting

C

Cathode – a negative electrode through which current flows out of an electrical device

Chemically clean – Clear of all traces of grease or impurities

Combustible – likely to catch fire, burn, or cause chemical reaction

Corrosion – disintegration of a material due to a chemical reaction

D

Dye sublimation – a printing process using specially designed inks with heat to transfer dye from one surface to another from a solid to a gas

E

Electrode – an electrical conductor used to make contact with part of an electrical circuit

Electrolyte – any substance containing free ions that will conduct electricity, also known as ionic solutions

Etch – a process using strong chemicals to cut into or corrode metal, in this case as part of the cleaning process

F

Fabrication – to construct something or put together

Findings – components used to hang or attach pieces of jewelry such as ear wires, catches, and jump rings

Forging – shaping metal by using localized compressive forces

G

Guillotine – a blade pulled down by a lever to cut sheet metal

J

Jig – a frame or clamp, in this case made from aluminum, used to hold and secure pieces of work during the anodizing process

L

Linocut printing – the surface of linoleum is cut away in parts to create an image that is then inked and printed onto the anodized aluminum

M

Malleability – the ability of a metal to be bent or shaped without breaking

Mandrel – a conical or tubular tool made in varying sizes from steel or wood used to create curved surfaces or circles

Microns – a millionth of a meter

Monoprinting – printing images in ink from one surface to another (printing plate to aluminum); suitable for printing one-of-a-kind images

O

Organic solvent – a liquid that is used to dissolve or thin a solid or liquid, such as nail-polish removers or glue thinners containing acetone, methyl acetate, and ethyl acetate

Oxide – a thin layer produced when an element or compound combines with oxygen, in this case aluminum oxide

P

Pickle – a mild acid solution (sometimes dilute sulphuric acid) used to clean oxide and residue off of metal, for example after soldering; in this case, it is used to remove dye from anodized aluminum

Pigment – a fine powder used as a colorant that is mixed with a binder to make a liquid (in this case water) or paste

Piercing – cutting metal with a jeweler's saw frame and blade to enable neat edges and fine detail

PNP (press-and-peel) paper – printed and used to create a relief pattern on the surface before or between dyeing

Pre-anodized aluminum – aluminum sheet that has undergone the anodizing process, been sealed airtight, and is ready to be dyed

Pressing – to form a shape in metal using a mold or die and applying pressure

Punch – a hardened steel or wood tool made for shaping, cutting, or indenting metal that is used in conjunction with a hammer or press

R

Resist – a solution or material used to block a surface from being altered

Rivet – a means of fastening layers of metal together, usually with wire or tubing

S

Silkscreen – a printing process suitable for multiple printing of the same image; a pattern is created in resist on silk stretched over a wooden frame, and ink is drawn over it, leaving an image on the surface below

Split mandrel – small drill bit with a split for holding emery paper

Stop out – to resist a dye (see resist)

T

Toner – ink used to print from a photocopier

Acknowledgments

I would like to say a huge thank you to all the artists featured in this book for allowing us to use images of their beautiful work.

Thanks to everyone at Lark Crafts for giving me this opportunity. Thanks especially to Marthe Le Van and Gavin Young for your faith, unfaltering enthusiasm, and patience. Thanks to David LaPlantz for your support and blessing. Thanks to the staff at UCA Farnham past and present: James Ward, Catherine Mayo, Richard Jones, Anita Bristow, Bex Burchell, Rebecca Skeels, Jonathon Jarvis, John Joyce, Nick Lott, Steve Whitehill, Marc Gray, Elroy Brown, Ivor Jones, and Abas Nazari. Thanks Rebecca Van der Rooijen at Bench peg; Angie Boyer at Craft and Design Magazine; Graham at Camberley Precision Sheet Metal; Trevor, Steve, and Nathan at HTS Anodizers; Karen at GT Camberley; Phil Richards at Clariant; Joel Gray at Becon Precision Engineering; everyone at Tom Foolery; and Nicki, Jed, Kim, Teresa, and Madeleine at Samson and Coles.

I'm grateful to all my friends and colleagues past and present at the Farnham Maltings and Arts Council South East for your support and general kicks up the backside when I doubted my abilities—Stephanie Barklam, Kate Martin, Bob Martin, Jane Friend, Pauline Smith, Dale Allen, SJB, the guys at Weller Designs, John Smith, Ruta Brown, members of the JSN, Gavin Stride, and especially my students.

To my extended family of friends, I thank you for everything. You mean the world to me Bex, Jen, Fi, Shev, Lyn, Suz Mul, Karina, Maxine, Jemma, Sammy, Nat, Si, Angela, Mark, Cat, Darren, Miss Bo, the Jacks, Cia, Sandys, Heidi, Tracey, Gav, Amy, Ben, Dai, Liz, and Hatty. Liz Hancock, my long-suffering studio mate, turns all my negatives to positives and thinks everything I do is great. Everyone needs a Liz! To Skeels, a massive thank you for all your inspirational input, pacifying, and brainstorms. I honestly couldn't have done it without you and your energy-filled enthusiasm.

To my family—Mum, Dad, Mark, Tracey, Charlotte, and Florence—thank you for always supporting me and never encouraging me to take the conventional path. Lastly to Mum, thank you for making me wear your handmade curtain clothes that I didn't appreciate at the time, and persevering with your persuasion that "different" is better than being one of the crowd.

About the Author

Clare Stiles designs and makes jewelry in anodized aluminum. From her studio in The Farnham Maltings in Surrey, England, she also undertakes commissions, designs prototypes, develops projects, teaches, and curates. Her jewelry is exhibited and sold at galleries and contemporary craft fairs throughout the United Kingdom and Ireland. In 2006, Clare received an Arts Council England Research Grant and a CIBAS Funding Award for Research. She is a member the Association of Contemporary Jewellers, a registered maker with the British Crafts Council, and a founder member of JSN (jewellers and silversmiths network). Clare has a degree from the Surrey Institute of Art and Design and attended the Reigate School of Art and Design, East Surrey College.

Index

Artist Index